Advance Praise for

THE BIG ONE

"A vital, compelling read. The next pandemic is inevitable. Preparation is our only option. And there is no one better to show how than the duo of Osterholm and Olshaker."

—Atul Gawande, MD, MPH, author of
Being Mortal and *The Checklist Manifesto*

"*The Big One* is an extraordinary book. It's part thrilling disaster movie script, part insightful history of public health, part personal story. It's also the best after-action report on Covid-19 yet, along with the best advice on how both individuals and government can survive the next one—and there will be a next one. It's a book you won't just read. You'll reread it and learn something new each time."

—John M. Barry, author of *The Great Influenza*

"If there was one person to turn to for future pandemic readiness, it would be Michael Osterholm, one of the world's leading epidemiologists. In *The Big One,* a landmark book, he, along with Mark Olshaker, provides a remarkably clear-eyed, science-driven, and comprehensive preventive approach."

—Eric Topol, MD, author of
Deep Medicine and *Super Agers*

"A riveting warning—and a road map. In *The Big One,* infectious disease expert Michael Osterholm and bestselling author Mark Olshaker deliver a gripping, scientifically grounded account of a plausible—and perhaps inevitable—global crisis...This book is a wake-up call we can't afford to ignore."

—Governor Michael O. Leavitt, former
US secretary of Health and Human Services

"Three pandemic viruses have swept across the globe in the last twenty years. There will be another pandemic. The only question is when and how severe. At once thrilling and haunting, *The Big One* provides a lifesaving guide for an inevitable event."

—Paul A. Offit, MD, author of
Tell Me When It's Over and *Autism's False Prophets*

"Osterholm is an invaluable expert whose tremendous knowledge and humility also make him a trustworthy guide. In *The Big One*, Osterholm and Olshaker take us through a tabletop exercise about a highly contagious, airborne virus capable of killing millions. We can only hope those in power heed its guidance."

—Ezekiel J. Emanuel, MD, PhD,
author of *Reinventing American Healthcare*

"With his unique blend of knowledge, communication skills, and Midwestern forthrightness, Osterholm has made himself indispensable in the fight to understand and overcome global infections. In *The Big One*, he marshals this depth and breadth to prepare us for the struggle that surely will come."

—Peter Hotez, MD, PhD, author of
The Deadly Rise of Anti-Science

"A clear-eyed examination of pandemic fault lines and what must be done to fix them...Osterholm and Olshaker weave a tale that at times borders on horror, interspersing the gripping story of the fictional virus with historical accounts of global pandemics and the experience of preparation and mitigation. But the story never veers too far into darkness without returning to practical advice."

—*Kirkus Reviews* (starred review)

"Osterholm and Olshaker shake us out of our post-Covid complacency to confront the potential for *The Big One*. Both a warning and a manual, the book provides clear and concise action plans for how we can collectively confront, prepare for, and respond to other, possibly worse pandemics. It also thoughtfully addresses the many ongoing debates about America's Covid response and shows how the American public may learn to trust again."

—Juliette Kayyem, former assistant secretary,
Department of Homeland Security

THE

BIG

ONE

THE
BIG
ONE

HOW WE MUST PREPARE FOR FUTURE
DEADLY PANDEMICS

MICHAEL T. OSTERHOLM, PhD, MPH
AND MARK OLSHAKER

LITTLE, BROWN SPARK

New York Boston London

Little, Brown Spark
Hachette Book Group
1290 Avenue of the Americas, New York, NY 10104
littlebrown.com

First Edition: September 2025

Little, Brown Spark is an imprint of Little, Brown and Company, a division of Hachette Book Group, Inc. The Little, Brown Spark name and logo are trademarks of Hachette Book Group, Inc.

The publisher is not responsible for websites (or their content) that are not owned by the publisher.

The Hachette Speakers Bureau provides a wide range of authors for speaking events. To find out more, go to hachettespeakersbureau.com or email hachettespeakers @hbgusa.com.

Little, Brown and Company books may be purchased in bulk for business, educational, or promotional use. For information, please contact your local bookseller or the Hachette Book Group Special Markets Department at special.markets @hbgusa.com.

Book interior design by Marie Mundaca

ISBN 9780316258340
LCCN 2025932647

1 2025

MRQ-T

Printed in Canada

To my beloved Fern

—Michael Osterholm

In memory of my father,
Bennett Olshaker, MD,
and
my father-in-law,
Louis J. Clemente, MD

—Mark Olshaker

Nothing was ready for the war which everyone expected.
       —Leo Tolstoy, *War and Peace*

For every complex problem there is an answer that is clear, simple, and wrong.
         —H. L. Mencken

CONTENTS

THE

BIG

ONE

PROLOGUE

What's past is prologue.
　　　　　　—William Shakespeare, *The Tempest*

It was the morning of December 30, 2019.

My partner, Fern, and I were on our way to the airport for a flight from Minneapolis–St. Paul to Spokane, where we would rent a car and drive to visit her family in Idaho for New Year's. We were heading down Interstate 494 when my cell phone rang. It was Lisa Schnirring, the news editor at CIDRAP—the Center for Infectious Disease Research and Policy at the University of Minnesota—which I founded in 2001 and direct.

"Mike," she began, without any introduction or her usual pleasantries, "ProMED is citing reports from Hong Kong and Taiwan about a mysterious pneumonia outbreak of unknown origin in Wuhan, China."[1] ProMED, the Program for Monitoring Emerging Diseases, which was established by the International Society for Infectious Diseases, at the time was one of the go-to online sources the public health community used for alerts about new and potentially significant infectious disease outbreaks.

"They're saying Chinese health officials are investigating," Lisa continued, "and the outbreak seems to be linked to a local seafood market."[2]

"Unknown origin" was not a phrase we liked to hear, and "seafood market" could mean a lot of different things. Asian food markets were often referred to as "wet" markets because of the continual melting of the ice that kept the fish fresh. But there was another meaning to "wet," since both domestically raised and wild animals, poultry, ducks, and geese were sold live and then slaughtered when purchased, a potential epidemiological nightmare of spraying bodily fluids, bloody ground, and cross-species germs mixing in the open air.

Every time epidemiologists or public health officers hear about an outbreak of unknown origin in some remote location, we begin mentally playing the odds, in the same way I imagine a seismologist does with a tremor reading, or a meteorologist with the National Hurricane Center does upon noting a tropical cyclone forming off the African coast. *Is there something to this, or is it just background noise? Is this something to alert the public to, or will it only cause needless worry and social disruption?*

Yet, always, in the backs of our minds: *Could this be the Big One?*, the pandemic that lurks as an ill-defined catastrophe in our fevered imaginations, threatening to test every scientific, governmental, and human resource we have. For people in my business, that prospect is at once too overwhelming to contemplate and too devastating not to.

The first thing we needed to figure out was whether the so-far-anecdotal reports were real, and whether or not they meant anything. Halfway around the world, and with so little specific information, it was hard to say. A lot of people in and out of the public health sector rely on CIDRAP News, and I never want us to be either caught flat-footed or guilty of crying wolf. Releasing sketchy information from CIDRAP before we knew the full story could create an echo-chamber effect, as other reporting agencies would pick up on it and spread the word. We needed to be careful not to contribute to the confusion before we had more data, but we didn't want to be behind the curve, either. Our credibility and, ultimately, public health's credibility were on the line.

"If you can confirm this from other sources, we should go with it," I said to Lisa. "Ask around and text me if you come up with anything. I'll call you when we get on the ground in Spokane."

"Right," she said. "I'll let you know."

As the plane rose through the clouds, I mulled over the possibilities. It was probably nothing to get excited about, I reassured myself. Pneumonia is pretty common, especially in older people. Often, physicians never identify the microbe causing the infection. In the medical realm, we're trained to consider common things over the more esoteric possibilities: When you hear hoofbeats, look for horses, not zebras, as the old saying goes.

Then again, all of us in public health are haunted by the memory of the cluster of cases of the rare *Pneumocystis carinii* pneumonia in Los Angeles and Kaposi sarcoma in New York that heralded the first American appearance of human immunodeficiency virus—HIV—in June 1981. As a Minnesota Department of Health epidemiologist, I happened to be at the CDC (Centers for Disease Control and Prevention) working on toxic shock syndrome the day Dr. James "Jim" Curran, head of the Sexually Transmitted Disease Control Division, who was to become one of the world's leading experts on HIV/AIDS, was assigned to investigate the first reports of that outbreak. Since Jim was working on a study of a new hepatitis B vaccine in gay men and I had been doing a study on how healthcare workers at a single hospital in Minnesota had become infected with hep B, he asked me to participate in a meeting with some of the leading lights in infectious disease at the CDC. At the time, none of us sitting around that table in the CDC director's conference room knew quite what to make of the data.

That meeting was certainly on my mind as I thought about Lisa's report. I was troubled by the possibility of a cluster of pneumonia cases and the mention of the live/wet market. The first indication of influenza outbreaks is often a pneumonia cluster, and with all the flu strains we were currently tracking, we really didn't want any of them to take off as H1N1 influenza virus had in 2009, or turn out to be something even worse. I was one of a small group that had studied pandemics most of my career and therefore had a pretty detailed sense of their potential to kill or harm both people and animals.

In 2005, in a highly unusual, coordinated effort with three major journals—the *New England Journal of Medicine, Nature,* and *Foreign Affairs*—I authored overview pieces for each publication on preparing for the next pandemic. Later, when Mark Olshaker and I published *Deadliest Enemy: Our War against Killer Germs* in 2017, we warned that a novel flu strain would be the most likely cause of a new pandemic on the order of the Great Influenza of 1918. We even included a fictional but realistic scenario showing how destructive and shattering such a disaster would be to all aspects of modern life.

In the course of this research and experience, one thing had become clear to me: You never know when a black swan event is going to turn into a red alert.

For me, that was the beginning of the Covid-19 pandemic, which turned out to be the most momentous, as well as the most painful, experience of my career in epidemiology so far, one that included both triumph and tragedy. We saw the development and production of a new class of vaccines that saved millions of lives. But we also knew that when the record of Covid-19 was finally compiled, there would be a lot to answer for: mistakes of judgment, denial of scientific evidence, overpromising and underperforming leadership, misleading or confusing communications, reliance on "experts" who actually weren't, inadequate drugs and supplies, insufficient healthcare facilities and personnel, intensified economic inequality, and institutions floundering to meet their stated missions. Some of these were understandable under the extraordinary circumstances. Some I still can't fathom.

It was understandable, for instance, that only a few of us in public health realized early that SARS-CoV-2, the virus that caused Covid-19, could be easily transmitted by infected individuals before they felt sick, or by those who never got symptoms. It was not understandable, on the other hand, that so many medical and public health officials were unable to consider or accept the fact that the virus was transmitted through the air across distances far greater than six feet from an infected person, or that a considerable percentage of that airborne transmission was microscopic aerosol particles rather than still tiny but heavier droplets expelled by sneezing or coughing that wouldn't float in the air very long or travel far.

Some of the problems we collectively encountered were structural, and not quick or easy to solve. But they all have to be addressed if we are to have a chance of triumphing in the microbe wars.

When the Covid pandemic began, we had no way of knowing that, accounting for reported and the estimated number of unreported cases, upward of 20 million people worldwide would die from the new virus within its first three years, to the point that the average life expectancy in even highly developed / high-income countries would be shortened by a matter of years; that many more scores of millions would need hospitalization or even greater intensive care; that many children would be orphaned and many families would lose their breadwinners; that many blue-collar and service industry workers would lose their jobs, their homes, and more, to the economic impacts that amplified existing socioeconomic inequalities to an exponential degree; that many cases of cancer, hypertension, cardiovascular events, and other diseases would go from preventable or treatable to potentially serious or fatal because people could not get access to early diagnosis or treatment; that many people would have their lives thrown off course by an unpredictable range of debilitating "long Covid" symptoms that would keep them from working and have profound emotional, sociological, and economic effects; that childhood vaccination programs and international efforts to curb other serious health threats such as malaria, HIV/AIDS, tuberculosis, cholera, dengue, Zika, and mpox (formerly called monkeypox) would be diminished or curtailed; that we would lose many healthcare workers—both directly from infection and indirectly from burnout and from young people deciding against entering medical professions at all; that educational progress would be set back, with many children, particularly in lower-income environments, losing several grades' worth of achievement; that global supply chains would be impacted, causing worldwide shortages and assembly-line shutdowns; that the entire perception of the workforce and office life would change, with widespread adoption of remote work; that the decline of brick-and-mortar retail and the rise of online sales would accelerate;

that supply and labor shortages and too much government economic stimulation would combine with Russia's invasion of, and protracted war with, Ukraine to trigger rampant global inflation; that the sociopolitical environment would become even more divisive in the United States and around the world, with vaccination, masking, and so-called lockdowns becoming proxy wars for partisan attitudes; and that trust in public health, the most valuable commodity in our arsenal, would erode to an alarming degree.

I've referred publicly to the Covid-19 pandemic as a microbial 9/11: a hinge in history where everything that comes after is radically different from everything that came before. But in a very real sense, as the preceding list illustrates, comparing the pandemic to that terrible and painful day in September 2001 does not do it "justice." In the United States, after the attacks—particularly in the first few days and months—people across the country largely came together against a common foe and to rebuild what was lost. We committed untold resources to anti-terrorism and created entirely new government departments. And through all the tragedy, we knew that terrorism could not bring our society to its knees as the terrorists had hoped and expected.

Covid-19 was able to accomplish that in a matter of weeks.

While some continue to doubt Covid's impact or importance, it is an undeniable fact, for example, that from March 2020 through October 2021, SARS-CoV-2 was the third-leading cause of death in the United States, behind only heart disease and cancer.[3] During that time, we learned that our modern-world system of international trade and travel doesn't work very well under extreme stress. For years after, the world struggled to recover from the initial impacts while simultaneously trying to adapt to a virus that refused to go away.

And yet, as bad and life-altering as that pandemic has been, it could have been worse. The Covid-19 virus killed 3.4 percent of those infected.[4] Imagine, for example, if the next pandemic had the ability to kill on an order of magnitude closer to those of the previous two serious coronavirus outbreaks: the 15 percent rate of SARS (severe acute respiratory syndrome),[5] first seen in 2002, or around 35 percent for MERS (Middle East respiratory syndrome),[6] experienced ten years later.

Fortunately, these two viruses were not highly infectious in most people, thus avoiding a potential pandemic scenario. But what if there was a virus with those mortality rates but with the much more dynamic transmission ability of the Omicron variant of SARS-CoV-2?

Such a terrifying prospect is what we call the Big One—the disaster, the catastrophe, the cataclysm, that haunts the midnight of every responsible epidemiologist's soul.

We've all heard repeated references to Covid-19 being the worst pandemic in a hundred years, as indeed it has been. But the one a century earlier was even worse, and a terrifying forecast of what we could be facing in the future.

The Great Influenza pandemic, which began in 1918 and circulated around the globe two and a half times by 1920, killed as many as 100 million people worldwide, far more than the blood-soaked world war that had recently ended.[7] The pandemic so altered the status quo that the global average life expectancy declined by a matter of years.[8] It is safe to say there was no one alive at the time who didn't know someone who succumbed to that terrible virus. And though it became known as the "Spanish flu" (because Spain was a noncombatant in World War I, it didn't censor its press and therefore reported case numbers openly), it is quite possible the Great Influenza first appeared in the unlikely location of Kansas.[9] There is also evidence it might have originated across the globe in Vietnam or China.[10] This uncertainty more than a century later shows how little we can be sure of in our ongoing war against deadly microbes.

As another frame of reference, HIV killed approximately 42 million people between 1981 and 2023,[11] the third year of the Covid-19 pandemic. As horrible as that is, it's not even close to the devastating potential of what an airborne-transmitted virus like a pandemic influenza strain or a new coronavirus could do in just a year or two.

You might conclude that with more than a century of scientific research and medical progress since 1918, we'd be in a far better position to combat a pandemic similar to the Spanish flu today. Sadly, we believe that assessment is built on a base of unfounded optimism. Compared with 1918, our modern world has almost four and a half times the population—ample human "wood" for the next viral wildfire; air travel that is possible between any two points on the planet within hours—far less time than a virus's incubation period; vast encroachment on forests and natural habitats that harbor animals that are viral reservoirs; hundreds of millions of humans and viral host animals, such as pigs, poultry, and cows, living cheek by jowl in agricultural settings that are viral genetic breeding farms; impoverished and densely packed tinderbox megacities without adequate hygienic facilities or medical support; more than a billion international border crossings each year; and globally interconnected supply chains dependent on just-in-time delivery systems, making the whole world's economy vulnerable to a disease that shuts down production in any one manufacturing region. Stated another way, humanity has become an extraordinarily efficient biological mixing bowl as well as a highly productive viral mutation factory. And on top of all these factors, the reality is that healthcare systems around the world are so broken—or, in some places, nonexistent—that we cannot adequately care for our populations in the best of times. Even in the United States, though Covid-19 was far less virulent than the 1918 flu, it still killed more people in the same amount of time—more than a million reported Covid deaths versus 675,000 for influenza.[12]

Clearly, Mother Nature still has the upper hand, and she is using many of the trappings of modern life to extend her reach, now against the backdrop of global climate change causing droughts that limit food and safe drinking water, and flooding that increases the transmission of vector-borne diseases such as malaria, dengue, and West Nile virus, spread by mosquitoes; and Lyme disease, spread by ticks.

This book is an attempt to reckon with our ever-more-perilous situation by someone with a prominent seat at the public health table from the very first days of the Covid-19 pandemic. Above all else, it is a story of conflicts: virus versus humanity, hubris versus humility, lofty vision versus dreams that can turn into nightmares. It is also about the mythology of what can be done in an ideal world versus the reality of what we can—and must—accomplish in the one we all live in. Its human characters include scientists and physicians, epidemiologists and public health officials, and politicians and media personalities, as well as millions of ordinary citizens around the world. Its microbial characters include influenza, smallpox, measles, cholera, Ebola, malaria, rabies, mpox, and the ever-increasing family of coronaviruses. It has other characters from the animal kingdom, too: bats, camels, and birds, to name just a few, play a critical role in harboring both influenza and coronaviruses.

Through all this, I stress humility, which is a strength, not a weakness. Indeed, it is an acknowledgment of reality. Bill Gates, who has had a major positive effect on global health in low-income regions through the work of the Gates Foundation (formerly the Bill & Melinda Gates Foundation), has quoted my epidemiologist friend Larry Brilliant: "Outbreaks are inevitable, but pandemics are optional." While I admire the aspirational nature of the statement, I have seen no evidence that it is remotely true when a readily transmitted respiratory virus with the "right" characteristics (i.e., wrong to humans and animals) spills over to humans in even one location in the world. In fact, quite the contrary. With all the tools, techniques, knowledge, and human resources available to us, we can limit some outbreaks, as we have done with SARS, MERS, and Ebola, but any thought that a pandemic caused by a highly contagious, airborne-transmissible respiratory virus—what we refer to as a "virus with wings"—capable of being spread by people who have no symptoms, with even moderate pathogenicity, and for which limited to no previous immunity exists in humans, can be halted before it spreads around much of the world is pie in the sky sprinkled with pixie dust.

Certain influenza viruses and coronaviruses are the poster children for the "virus with wings" designation, though most have not previously caused a pandemic for reasons that are unclear. For example, SARS and MERS have caused deadly outbreaks around the world since 2003, but they were limited in the number of cases and relatively quickly brought under control because these viruses lacked the infectiousness of SARS-Cov-2. Despite the eighteen different H (hemagglutinin) types of influenza A, only H1, H2, and H3 have caused pandemics in the past hundred-plus years.[13] Whether other influenza H-type viruses (like H5 or H7) will cause a future pandemic is unknown.

Another family of viruses — filoviruses — remains a big question mark with regard to pandemic potential. Ebola and Marburg, two of the most recognized diseases in this family, have been major public health concerns, as outbreaks have increased in number and size over the past twenty years. The illnesses are characterized as hemorrhagic fevers, with case fatality rates of 50 to 85 percent.[14] The largest filovirus outbreak, from Ebola, occurred in West Africa and ran from July 2014 to June 2016. More than 28,000 people were infected and at least 11,300 died.[15] Contact with body fluid of an infected person was considered the primary mode of transmission.

In the middle of that epidemic, I led an international group of twenty senior experts in Ebola virology and epidemiology to review what we knew about the virus and how it was transmitted. We summarized our review in *mBio,* a major infectious disease journal. We concluded that most cases did result from contact with the body fluids of infected individuals. However, after an exhaustive review of all the available outbreak investigation reports and laboratory research, we hypothesized that Ebola viruses have the potential to be respiratory pathogens; in other words, transmitted by traveling through the air.[16] We received some serious negative pushback from the infectious disease community on our conclusions regarding respiratory transmission potential. None of the critics had any data to refute our conclusion; frankly, they just couldn't accept that an airborne Ebola transmission might ever occur. Nothing has changed my scientific opinion since that 2015 publication that respiratory transmission might very well occur in a future outbreak in such a way to make a filovirus a potential virus with wings. I surely hope not, but we must be prepared for that possibility. The world has virtually no existing immunity to the filoviruses, so significant airborne transmission potential would put us in big trouble.

There are other respiratory-transmitted viruses that are highly infectious and might be capable of causing a pandemic. Measles is a good example. It is as infectious as influenza viruses and coronaviruses, but because it has undergone few mutations in the past fifty to sixty years,[17] human immunity from a previous measles exposure, whether it be from infection or vaccination, is able to protect you. A similar situation exists with a number of other viral agents.

Of note, I don't believe bacteria meet the qualification for a separate "bacteria with wings" category. Pneumonic plague, caused by the bacteria *Yersinia pestis,* was the scourge of the Middle Ages' Black Death. However, today, the combination of effective antibiotics and rodent- and flea-control programs[18] means it poses almost no risk of causing a modern pandemic. This is not to rule out the possibility of something unexpected or unknown emerging, but the cards are significantly stacked against that happening with bacteria.

But rapidly spreading respiratory virus–caused pandemics are as much a fact of life as war and crime. All we can do is our best to mitigate their effects and shorten their duration and spread.

It has often been said that in any conflict, we start off by fighting the last war. While this is sometimes leveled as a criticism, there is a reason military leaders and strategists study past battles and campaigns. Only by analyzing and understanding how previous approaches played out—what worked and what didn't, which assumptions about the enemy turned out to be valid and which did not—can our fighting forces improve their preparation for the next confrontation.

Specific plans may go out the window at the first encounter with the enemy. But being ready for a range of eventualities would seem to be a critical exercise. General Dwight D. Eisenhower may have put it best and most succinctly when he declared to a military conference in 1957, "In preparing for battle, I have always found that plans are useless, but planning is indispensable."

One of the ways government and public health officials try to prepare for pandemics and other natural and man-made disasters is through tabletop simulation exercises in which all the high-level participants must make critical split-second decisions, usually with incomplete information. I have participated in these exercises as both the leader who narrates the situation and series of events and a player making the decisions. Each of our chapters begins with such an episode from such a scenario.

In an attempt to show what it would take to prepare for the inevitable Big One, we have imagined another coronavirus pandemic, but with much higher virulence and transmissibility. Our thought experiment is the equivalent of a tabletop simulation exercise and has been vetted by experts on every relevant aspect. With this scientifically plausible and very possible exercise, we really could be staring at the Big One. Because such a pandemic would necessarily involve so many people and places, with so many moving parts, our thought experiment features a large cast of characters located around the world from all economic, social, and political strata — just as we'd see in real life. While it may seem overwhelming at first (again, as it would in an actual new outbreak), some characters will appear throughout multiple chapters as we go through the pandemic's storyline. Each one plays a role in helping us illustrate and develop the key takeaways in the story. We have included a "Thought Experiment Cast of Characters" as an appendix to make it easier to keep track of everyone.

We're all familiar with the science fiction and postapocalyptic stories, movies, and television shows that posit fatality rates of 75 or 90 percent, leaving only a handful of intrepid survivors to carry on and rebuild society. That, we can say, fortunately, falls in the realm of apocalyptic fiction. No infectious disease in human history has ever inflicted such devastation. But the reality is devastating enough.

For the sake of realism in our thought experiment, we have assumed that no major scientific or public health initiatives will be taken in advance of the pandemic, other than 2023's declared US government's Project NextGen, which aims to develop better coronavirus vaccines and antiviral drugs, and the National Institutes of

Health (NIH) support for improved influenza vaccine development. While we certainly hope those efforts provide early, critical support to find better vaccines and drugs, we don't believe the $5 billion investment and initial eighteen-month time frame are anything more than a down payment on the research and development we will need in order to accomplish this very important goal. Accordingly, our thought experiment will show the consequences of being essentially reactive rather than proactive in defending against lurking microbial threats — in particular, the sort that will make the pain and suffering from the Covid pandemic seem like a mere warning shot across the bow.

In progressive fashion, each chapter will deal with events and topics highlighted in that chapter's fictional opening scenario. Though there will be some overlap due to the complex interrelatedness of so many of the issues, we have presented topics in the order that government, public health, and the public at large would have to face and deal with them when the Big One hits. We begin with an understanding of the basic science as a foundation in Chapter One and go from there to airborne transmission — a reality we faced with Covid and certainly will face with the Big One — in Chapter Two. The next chapters delve into the substance of how we will attempt to manage the pandemic: Chapter Three discusses mandates and other nonpharmaceutical mitigation efforts, Chapter Four addresses medical countermeasures, Chapter Five describes what we will need in terms of effective communication, and Chapter Six explains disease surveillance. Chapter Seven addresses policy and politics, and we conclude in Chapter Eight by discussing where we go from here.

In each chapter, we propose what can and should be done to prepare in terms of the topic under consideration, while at the same time calling out the simplistic notions that caused so much misinformation, distrust, and unnecessary suffering and deaths during the Covid pandemic, as they will do nothing to help us fight the Big One when it comes.

It is difficult for most of us to think in abstract terms about a vague threat sometime in the indeterminant future. But that is exactly what the military does, because its leaders know the consequences of not doing so are almost unimaginable. Failing to prepare for the next pandemic will surely mean the loss of countless lives, which will make any investment made beforehand seem insignificant when compared with the economic devastation unpreparedness will wreak.

Prior to Covid-19, we believed the next pandemic would be caused by another influenza strain. Even after the appearance of SARS and MERS, we didn't think coronaviruses had the capacity to transmit and infect as efficiently as SARS-CoV-2 did, which gives us another compelling reason for humility as we collectively approach this subject. Now we realize that coronaviruses also have that potential.

At this point, we still see influenza and coronaviruses as the two viruses most likely to cause a pandemic on the order of the 1918 Great Influenza. But there are probably others lurking, possibly in a bat cave or on a farm where pigs and poultry are raised in close proximity.

Viruses can be maddeningly and chillingly unpredictable. The flavivirus Zika was first detected in a rhesus monkey in Uganda's Zika Forest in 1947, and for more than six decades it was known to cause only a pink rash, conjunctivitis, occasional joint and muscle pain, or no symptoms at all.[19] By 2015, however, it was triggering the paralyzing autoimmune Guillain-Barré syndrome,[20] and by the time it reached Brazil, it was causing babies of infected mothers to be born with microcephaly[21]—smaller than normal heads and brains that do not develop properly. By the same token, before SARS, coronaviruses produced mild to moderate respiratory illnesses in humans. And within twenty years of SARS's first appearance, we saw SARS-CoV-2 collectively kill millions.

Let us face the sober reality: We spend many billions of dollars every year on national defense and security in the United States, but pandemics have killed more human beings in modern times than all the wars in history. It is no exaggeration to say that each of us remains in far greater constant danger from microbial enemies than from human ones. What will it take to convince us that, short of a thermonuclear war or the slow-moving tsunami of climate change, the greatest threat we face to our national security and way of life is an invasion of deadly microbes?

We may find ourselves needing that planning far sooner than we expect. The time to start is now.

CHAPTER ONE

VIRUS VERSUS
HUMANS

The single biggest threat to man's continued dominance on the planet is the virus.

—Joshua Lederberg, MD

[Pandemic Day 1]

For several years, Warsame Amir Osman farmed a small plot of land on the edge of Lag Badana National Park, near the southern tip of Somalia, close to the border with Kenya. He had worked other locations, but with severe drought lasting more than ten years, those lands would no longer bear enough crops to feed his family or to sell for a little money or provide vegetation for his two camels.

This land, too, was quickly becoming barren. Still, Warsame and his three sons worked almost every day, until the morning he woke up with his head and muscles aching, with chills and a cough that would not clear. Just walking left him breathless.

The next morning he felt worse. And now his middle son, Cabdi, was also coughing.

Warsame's wife, Yasmiin, went to find Jamilah Shamshi, the community health worker. There was little Shamshi could do beyond offering fluids, food, and palliative care. She was used to dealing with cholera, measles, chikungunya, and dengue, and it was common for her not to see people until they were in desperate condition. Some recovered under her care. If they didn't, she held their hand at the end and tried to comfort their survivors.

———————

Not far away, Axlam Omar Yussef and his wife, Zahi, had decided to leave their village to find a more stable life, with more dependable sources of food and water. They set out on foot, carrying their one-year-old daughter, Hani, and three-year-old son, Mohamed, their camel bearing their few possessions. It would take a week or more to reach the Hagadera Refugee Camp in Kenya, but they'd heard there were opportunities in Nairobi for those willing to work hard. At least in Hagadera, maybe, their children would have enough to eat.

During the journey, Zahi fell ill, with a hacking cough that made walking difficult. Hani was also coughing, sometimes gasping for air, but they had no option but to continue.

———————

In the small port town of Burgabo, bordering the northern coastline of Lag Badana National Park, Fawzia Noor Mahad and her husband, Nadifa, had eked out a modest living preparing charcoal to be shipped overseas, until the UN Security Council banned the export because it profited Al Shabaab, which the UN had labeled a terrorist group affiliated with Al Qaeda. Nadifa tried farming another man's land, but Al Shabaab took half the crops. When Nadifa and his employer asked to keep a little more to feed their families, the guerrillas killed them as a lesson to others, leaving Fawzia alone and pregnant.

As the army of Somalia and the Kenya Defence Forces fought to drive Al Shabaab out of Burgabo, Fawzia knew she was not safe. She set out to seek sanctuary in the Hagadera Refugee Camp.

———————

A few days later, at the northern edge of Lag Badana, Jamilah Shamshi, the community health worker, was seeing many more people with the same symptoms as farmer Warsame Amir Osman and his son had. Some were not too sick, but several, including young children, were having trouble breathing and were running high fevers.

Jamilah didn't know what this disease was, but it was making enough people seriously ill that she reported it to the region's mobile health worker. He could contact the World Health Organization (WHO) office in Mogadishu, reaching doctors with more training and medicine.

———

Dr. Daniel Onyango was chief health manager for the International Rescue Committee at the Hagadera camp, overseeing a 140-bed hospital with a 15-bed isolation facility and four additional health posts throughout the camp. Onyango had a medical degree from Kenyatta University School of Health Sciences and could have had a profitable practice in Nairobi treating the well-off. But, heavily influenced as a teen reading Jean-Paul Sartre and Frantz Fanon, he'd become a physician to try to better the lives of those the world had forgotten, the ones Fanon called "the wretched of the earth."

Hagadera is a sprawling, makeshift city of wooden buildings, tents, and handmade huts of sticks, cloth, plastic, and anything else residents can scrounge. It sits at the southern end of a three-camp complex known as Dadaab that sprang up in the early 1990s and continued to grow over the next few decades as hundreds of thousands of displaced persons fled armed conflict and food insecurity in Somalia, Ethiopia, Sudan, Burundi, Rwanda, and the Democratic Republic of Congo. Dadaab has better access to drinkable water than most of the region, and that is likely why in 1941 the British built a fort on the site, which is now the Hagadera camp. Still, water-borne disease outbreaks, especially cholera, happen with some frequency. Attempts to close the camps have failed due to the ongoing need and inability to repatriate its residents for long—or to send them anywhere else. The constant influx of new refugees—at Hagadera, almost exclusively Somalis—has made Dadaab Kenya's third-largest population center, after Nairobi and Mombasa. It is also an economic hub, as the A3 highway, which links Nairobi with Somalia, offers easy access, making the area a key destination for livestock trade.

To international relief agencies, camp inhabitants can sadly merge into overwhelming statistics. To Onyango, who saw in each patient a unique human being, there was nonetheless a depressing sameness to their stories: the struggle for survival and the need to escape—from hunger and thirst; economic collapse; and the threat, or actual experience, of pillaging, rape, and torture from terrorists. So many refugees arrived severely malnourished, particularly the children. Last year, thirty-two babies died at the hospital from malnourishment and related conditions—a number Onyango knew vastly underestimated reality, as so many died en route to or within the camp but outside the hospital.

Along with food insecurity, cholera, spread through contaminated water, is a constant peril, and the severe diarrhea and vomiting it causes can lead to dehydration and death within hours. Children, weakened by malnutrition and lack of medical care, are most vulnerable. For them, a measles infection can easily lead to fatal diarrhea, pneumonia, and encephalitis.

Then there is HIV/AIDS, tuberculosis, diabetes, and hypertension. While any one of these alone is life altering, when they are combined with the acute malnutrition, fear of violence, and homelessness the refugees faced, Onyango thought it no wonder that so many suffered from post-traumatic stress, chronic depression, and other severe mental health problems. He treated their symptoms as best he could, but there was precious little that he could do.

———

It was Wednesday morning when Dr. Onyango saw Hani Axlam Omar, a one-year-old who had arrived in camp the day before with her parents and sibling, a boy of three. All were undernourished, but the baby was in the worst shape, with trouble breathing, diarrhea, and a nonproductive cough. The boy presented with the same symptoms, but not as severe. The parents had been sick a few days earlier but now seemed okay.

The doctor swabbed the children's nasal passages to test for SARS-CoV-2, the virus that causes Covid-19, but the results were negative. He started both on IVs and an injection of antibiotics. Onyango wasn't sure what they had, but he told the nurses to watch the baby, in particular, throughout the night.

The next morning, Mohamed was stable, but Hani was worse. Her distraught mother sat by her bedside all day, feeling utterly helpless.

On her third day of hospitalization, Hani died in her mother's arms. Dr. Onyango had seen too many babies and young children die, but there was no recourse but to keep going.

That same day, Fawzia Noor Mahad, another recent arrival from Somalia, was brought to the hospital. Her symptoms were similar to Hani and Mohamed's and she also tested negative for Covid. She said she started feeling poorly during the trek from Burgabo but thought it was from pregnancy and the long journey, made more difficult because her camel was moving more slowly than usual. The many camels in Somalia and Kenya are highly prized for their milk, meat, and resilience — able to go a month or more in drought conditions and still produce milk. Most rural folk know instinctively if they are sick.

Onyango wondered, if a camel and its owner were both sick, could this be MERS, the Middle East respiratory syndrome? A coronavirus that first appeared in the Arabian Peninsula in 2012 and was usually carried through camels, MERS fortunately did not spread easily to humans. But while Covid had a fatality rate around 3.4 percent, the MERS death rate was about one in three.

To Onyango's alarm, within hours, several more patients with the same symptoms were brought to the hospital.

Dr. Joseph Ndembi was chief epidemiologist for the Eastern Africa Regional Coordination Centre (EA-RCC), headquartered at Kenyatta National Hospital in Nairobi. He was in his office when Dr. Onyango called. Onyango had followed Ndembi in medical school, and Ndembi held the younger man in high regard. Calls about a disease outbreak at a refugee camp were, unfortunately, not uncommon. But Onyango seemed particularly urgent. He had seen seven individuals in the last three days with similar symptoms: respiratory distress, fever, chills, and body aches. Some also had diarrhea, and a one-year-old girl had died. He'd tested for and/or ruled out all of the obvious things, including Covid. Onyango added that one patient's camel had been sick, which concerned both men. Ndembi said he would send a team to investigate.

About an hour later, Ndembi took a call from Dr. Benjamin Amayana, a physician in Nairobi's Eastleigh suburb, known as "Little Mogadishu" for its predominantly Somali population and bustling economic pace. Dr. Amayana reported a cluster of SARDS — sudden acute respiratory distress syndrome — that sounded suspiciously like what Daniel Onyango had described, though no one in Eastleigh had died.

Eastleigh receives daily traffic from Hagadera, with most people making the seven-and-a-half-hour trip on hot, crowded buses. It was probably no coincidence that Ndembi received these two calls on the same day. Epidemiologists had worried about MERS making its way down the Arabian Peninsula through Yemen to Eretria, Djibouti, and Ethiopia, into Somalia and Kenya. Could this be what they were seeing? Please God, no, *Ndembi said to himself.*

Ndembi contacted the Kenyan National Virology Reference Laboratory, preparing them to receive blood samples and nasal and throat swabs his teams would collect from people and possible reservoir animals in the areas affected. They had no postmortem tissue from baby Hani because her family followed the local practice of rapid burial. Ndembi also contacted the Africa Centres for Disease Control to see if they'd gotten reports of the illness showing up elsewhere. Both the Kenya country office and the WHO Regional Office for Africa (AFRO) would be notified, then the Africa CDC regional center. He knew they'd need a strategy in case the illness spread further. Based on how rapidly the suspected case numbers were growing, that seemed likely.

The next day, Ndembi assembled two teams that set out in white Toyota Land Cruisers supplied by the UN, one to Eastleigh and the other to Hagadera, to interview anyone who had the described symptoms, as well as family members and close contacts, sick or not. He'd already coordinated to get whatever resources he might need from EA-RCC.

The teams combed their assigned areas, speaking with everyone Drs. Onyango and Amayana referred them to, gathering information about all the contacts they'd had since they or anyone in their immediate family had started feeling ill. Many had recently been in Somalia, and they'd either been sick or knew of people who were sick there. The teams took blood samples and nasal and throat swabs from all the contacts they could find — a challenge, given that most of the ill had traveled recently, in many cases sharing a bus or cramped encampment with equally transient strangers.

Of particular interest were people who'd had the mystery disease and recovered, such as Axlam Omar Yussef and his wife. It was too early to isolate specific antibodies, since whatever was making people sick hadn't yet been identified, but their blood might shed light on what they were infected with. Staff collected their samples as the parents stood watch over Mohamed, still in the hospital. The team learned that two staff members who had tended to Hani were now ill, but with the high patient load, both kept working. The team took samples from them, their families, and anyone else they'd been in contact with.

Each sample was carefully labeled and recorded before being taken to Kenya's National Public Health Laboratories. The first step was to run a research polymerase chain reaction (PCR) test for evidence of any type of coronavirus infection, though the test would not identify the specific type of the virus. Some samples were placed on Vero cells, a standard cell line with a population of cells that can repeatedly be grown and maintained in a laboratory culture for extended periods over which microbes can grow. Others went onto petri plates to check for bacteria.

The nonspecific coronavirus PCR test results came back positive within hours. It took two more days for the virus to grow on the Vero cells, but six days after the first samples had arrived in the lab, technicians were able to identify the mystery pathogen. It was a novel coronavirus—not SARS, MERS, or SARS-CoV-2.

When the genetic sequencing was completed two days later, the lab finding was confirmed—fourteen days after Onyango first saw baby Hani at the Hagadera hospital: This was a coronavirus no one had seen before.

To Ndembi and the lab staff, the results were like a biological bomb going off.

The technicians stood silently for several moments, not making eye contact; everyone concentrated their gaze on the genetic profiles. Finally, Ndembi said, "We'd better get Adamu in here."

Dr. Adamu Kimani, director of Kenya's National Virology Reference Lab, arrived within minutes. He stared at the viral sequence data. "It's confirmed, then," he stated grimly. "We are dealing with a new coronavirus. We know it's highly transmissible, and it's spreading."

Kimani picked up his phone. "Stay with me while I call WHO and AFRO," he said to Ndembi. He paused, turning away to look out a window. "I pray I'm wrong, but we may be looking at a world about to change."

———

The lab published the genetic sequence of the novel coronavirus on the website of GISAID—the international repository of influenza and coronavirus genetic data. On a call to Dr. Jeremy Davies, executive director of the WHO's Health Emergencies Programme in Geneva, Kimani briefed WHO leadership on the number of suspected cases they'd recorded so far and how rapidly the virus was spreading. Kimani urged them to alert governments and the public health establishment throughout the world about the new "pathogen of concern."

WHO published information on the new coronavirus in Disease Outbreak News, including a description of presumed cases in Kenya and Somalia, as well as the genetic sequence. A notice went out internationally to doctors and emergency departments with a description of the symptoms, requesting that they report to their local health department anyone meeting the case definition who had been in either Kenya or Somalia, or had been in contact with someone who had. An incident management support team (IMST) was activated to coordinate WHO assistance from headquarters and across regional and country levels, and headquarters set up an international video call with IMST members, WHO regional offices, representatives from the EA-RCC, the European Centre for Disease Prevention and Control (ECDC), and the US CDC, as well as leading academic institutions worldwide. WHO had been criticized for what was perceived as a slow and poorly coordinated response to Covid-19. If this new coronavirus turned out to be as great a threat as field reports seemed to indicate, Dr. Davies and WHO director-general Dr. Kolawole Adebayo did not want the organization to be caught flat-footed again. They called a meeting of the Public Health Emergency of International Concern Committee.

Response to publication of the new coronavirus's genetic sequence and WHO's notice was swift. Calls and emails flooded in from concerned hospitals, doctors, and health departments all over the world—many of whom had suspect cases—as well as the media. Everyone wanted to know what was being done to prevent spread of the new virus and how people should protect themselves. Without knowing more than what had already been shared, WHO released a statement that investigation was ongoing and advising healthcare facilities that anyone working with suspected cases should wear a surgical mask, unless they were doing an aerosol-generating procedure, in which case an N95 respirator was indicated.

————————

Etienne Navarre, a twenty-six-year-old aid worker for an NGO (nongovernmental organization) was returning home to Paris after volunteering at the

Hagadera camp. He departed from Nairobi's Jomo Kenyatta International Airport, the busiest international air hub in East Africa, on a six-and-a-half-hour flight to Istanbul, one of the ten busiest airports in the world. After a three-hour layover, during which he got something to eat and used the men's room, Etienne boarded another plane for the three-hour-and-forty-five-minute flight to Paris's Charles de Gaulle Airport.

It was on the first flight that he started coughing and feeling run-down, which he attributed to the long hours and demanding work at the camp. He'd hoped that getting a proper meal in Istanbul would make him feel better, but on the second flight, Etienne felt even worse. His chest felt heavy, and his muscles ached.

He would have to see a doctor when he got home.

Taramin Wenda was in the import/export business in his hometown of Jakarta, Indonesia. He had spent several weeks in Kenya working on deals for various commodities, most importantly tea and honey. After a few days of meeting with contacts in Nairobi, he flew to Wajir, in the northeastern part of the country, not far from the border with Somalia. There he met with an associate who, wanting him to invest in the camel business, took him to the town's biggest camel auction. Wajir is known as Kenya's livestock capital, with the largest camel population in the country. Taramin said he would consider it, but the other business was his priority.

After a few days in Wajir, he returned to Nairobi for some additional business before beginning the trip home with a flight to Istanbul. He had a five-hour-and-ten-minute layover there, but since he was flying business class, he could spend it in the airline lounge, where he had a shower and hot meal before settling down in the expansive relaxation area. Taramin always enjoyed using such time to meet people from all over. He talked to at least ten, networking with four potential new business leads.

Taramin slept for most of the eleven-and-a-half-hour flight to Jakarta, so when he got home, he was ready to start another busy day.

Curt Ashworth had spent his college semester break volunteering with his church's international outreach program in Kenya. He was eager now to get home to see his family in Georgia before returning to school. His father was a colonel in the Marines, assigned to the Marine Corps Logistics Base in Albany. A military brat, Curt had lived throughout the United States and overseas, which had encouraged his interest in international relations.

His eight-and-a-half-hour flight from Nairobi to Frankfurt was packed, and several times he had to wait in line for the economy-class lavatory. He arrived in Germany at 5:30 a.m., so most of the airport shops were closed. He hadn't gotten any sleep on the flight, but in the early-morning quiet, he found a seating area where he could stretch out. He crashed almost instantly and slept until his phone alarm went off four hours later, just before his flight was called.

The next leg of his journey—almost as long as the first—was on a crowded flight to Hartsfield International in Atlanta. By the time it landed, Curt was stiff from sitting, his joints ached, and he felt feverish. His throat was scratchy, too, which he attributed to the dry cabin air.

He was delighted when he exited International Arrivals to see his parents, who were holding a "Welcome Home" sign. They'd figured he would be exhausted after almost twenty-four hours of traveling and had booked a hotel room nearby so they could all rest before the three-hour drive home. They let him sleep until the hotel's breakfast service was almost over. Even then, there were lines at the well-stocked buffet.

Abdirahim Ali Salat was a young man with ambition. Born in a rural village in eastern Somalia, he had worked hard and been accepted at the university in the port city of Kismayo, along the southern Somali coast. In Kismayo, Abdirahim had met his wife, Bishaaro. They had three children, a girl and two boys, ranging from six months to almost four years old. He'd attained a bachelor's degree in business administration from Kismayo University and a diploma in information technology from the Kismayo Institute of Research and Community Development. There were jobs in Kismayo, but Abdirahim didn't feel he could reach his full potential there, especially amid the constant threat from Al Shabaab.

His older brother Abshir had emigrated to the United States, settling in a large Somali community in Minneapolis, Minnesota. Abshir was adept with computers and had gotten several certifications after taking courses online, so even without a college degree, he'd found a good job in tech support. He told

Abdirahim there would be opportunities for him in America, in a much safer environment for his children, and he could live with Abshir's family until he got on his feet. Abdirahim and Bishaaro agreed that once he was settled in Minnesota, had a job, and saved some money, he would send for her and the children to start a new life in the United States.

The brother of a fellow university student had a delivery truck, and Abdirahim paid him to drive three hundred miles up the coast to Mogadishu's Aden Adde International Airport for the four-and-a-half-hour flight to Hamad International Airport in Doha, Qatar. During a seven-and-a-half-hour layover, Abdirahim relaxed in a food court.

On the second leg of his journey, a sixteen-and-a-half-hour flight to Dallas Fort Worth International Airport in Texas, Abdirahim started feeling sick. His throat was scratchy, and he felt feverish.

He cleared US Customs and Immigration during a layover of more than eight hours in Dallas Fort Worth. His chest was tight, and his muscles felt weak, but he did his best to hide it, not wanting any issues with Immigration or to be stranded in Texas, where he knew no one. He just needed to get to Minnesota, go home with Abshir, and rest.

Midway through his two-and-a-half-hour flight to Minneapolis–St. Paul, Abdirahim was sweating profusely, and he couldn't stop coughing. A flight attendant asked if he was all right. He told her that he was, though he wasn't sure that was true. She brought him tea, which he drank more to convince her he was okay than because he wanted it.

When Abdirahim emerged from the arrival gate, Abshir was shocked by his brother's appearance. Abshir took him home to rest. The next day, after sleeping fourteen hours, Abdirahim was worse. Abshir insisted on taking him to the local hospital's emergency room.

Abshir went to the reception desk to check him in, and Abdirahim found one of the few unoccupied chairs in the waiting area. As in most US hospitals, although it was far better than what Abdirahim would have experienced back home, the emergency department was stretched thin. It was more than an hour before the triage nurse saw him. As soon as she did, she gave him a paper mask to wear and immediately fetched Dr. Erin Thomas, the attending physician in charge of the shift.

Once she had the full picture of Abdirahim's trip and his symptoms, Dr. Thomas instructed the nurse, "Let's get him in an isolation room while we

run tests. And let's all follow the Precautionary Principle. I want everyone in
contact with him wearing an N95 respirator until we know what this is."

MICROBES WERE HERE LONG before us, and probably will be here long
after us. As a species, we reproduce on average every twenty-five years,
the rough definition of a human generation. Microbes create a new
generation in anywhere from six to ten minutes (as with *Clostridium*
perfringens, a bacterium that causes acute gastrointestinal infection in
humans) to as long as fifteen to twenty hours (as with *Mycobacterium*
tuberculosis, which causes tuberculosis). In the case of SARS-CoV-2, a
new generation is produced in about ten hours.[1] This gives the *Clos-*
tridium microbe an adaptive advantage over humans of as much
as 40 million to 1: 40 million microbial generations to 1 human.
Covid's advantage was a mere 220,000 to 1.

In the eons since our mutual single-cell ancestors emerged from
the same primordial organic soup and evolved along divergent but
parallel paths, microbes and humans have both been developing
ways to deal with each other: laissez-faire indifference, mutual
dependency, and outright hostility.

Thankfully, most microbes live in harmony with us. Some simply
leave us alone and have little impact on our lives, and others are critical to our health and well-being, such as the trillions and trillions of
bacteria in our gut that not only facilitate digestion and nutrition
but also impact our immune system and general homeostasis — the
process by which living organisms maintain internal stability.

These benign or even helpful microbes are not the epidemiologists' quarry. Rather, we're after the dangerous microbes — the ones
that cause disease and death and have evolved and adapted to prey
on us. The battle lines are clear: the microbes' genetic simplicity and
evolutionary flexibility against our intellect, creativity, collective
social cooperation, and political will.

We tend to anthropomorphize animals and things that would do
us harm. While it may be appropriate to assign malevolent motives
to human beings, trying to do so for anything below us in the evolutionary order isn't useful or meaningful. A lion stalking an antelope

on the Serengeti is certainly predatory, but it doesn't hate that antelope or want to avenge some perceived wrong. It simply needs to eat to survive. That is its nature.

By the same token, a "bad" microbe isn't out to do us harm. Its mission in life is to reproduce and survive, and for that it needs a compatible host environment. If it happens to damage or hurt that environment, well, that is just nature in action. If it is deadly to too many of its hosts, however, it will die off, so some balance is in the microbe's best interest. The reality is that we humans have far more negative intent when we swat an annoying fly or spray a mosquito than a microbe has when it's taking over a host cell.

But malicious or not, these tiny predators are everywhere. And their ubiquity is part of what makes them such a formidable adversary. Microbes are governed entirely by self-interest—relentless self-interest.

We refer to the vast array of these and other tiny "bugs" as the microbiome. So pervasive are microbes that, despite their tiny size, the earth's microbiome outweighs every other component of the planet's biomass—all other life-forms—combined! There are numerous families and orders of magnitude within the earth's microbiome, including, from the smallest and least complex on up: prions, viruses, rickettsia, bacteria, fungi, and parasites. For our purposes in focusing on epidemics (localized or regional outbreaks) and pandemics (global), the two that most concern us are bacteria and viruses.

Bacteria are complete, self-contained entities, capable of respiration, movement, taking in nourishment and eliminating waste, defending against enemies, and, most important, reproduction.

Viruses, on the other hand, are even tinier microscopic entities. Unlike bacteria, they are not, strictly speaking, alive. But we can't say they are lifeless or inorganic, either. They exist in a netherworld between life and lifeless. A virus is something like a microbial version of Frankenstein's monster: It can't bring itself to life, but it can keep going by attaching itself to other organic forms to survive and reproduce.

Indeed, replication is a virus's only goal. While there's no sense in assigning human-like motives to microbes, we can say that viruses are always "looking" for cells they can penetrate and whose reproductive mechanism they can hijack. Once they achieve that, they begin churning out as many copies of themselves as possible, in the countless millions. That may lead to an individual's common cold or represent the opening salvo in a pandemic.

To be sure, we are not helpless against this marauding horde. Among the weapons in our arsenal are antimicrobials—chemical compounds, either occurring in nature or scientifically synthesized, that kill or inhibit the growth of various forms of viruses, bacteria, fungi, and parasites. Antimicrobials are our Excalibur in the fight against microbial infections. But this sword, alas, is growing blunter.

The world is facing a crisis of antimicrobial resistance that is only growing worse. Take antibiotics, the antimicrobials that target bacteria, for example. It is an unavoidable reality that the more antibiotics are used, the more opportunity there is for bacteria to develop resistance. Basically, while exposure to the right antibiotic will neutralize most of the targeted microbes, a small percentage could, due to a random mutation, develop immunity. As that mutation reproduces—and remember, they reproduce quickly—these immune bacteria can take over the strain, crowding out those less resistant. Many previously potent antibiotics have been rendered less effective against various bacteria, and some are completely ineffective. The truth is, we are losing powerful antibiotics faster than we are developing new ones. If political and scientific leadership don't take decisive steps to address this challenge globally, we could plunge back into a pre-antibiotic era where, for example, bacterial diseases like staphylococcus (staph) and streptococcus (strep) infections, as well as tuberculosis, could wreak havoc on previously healthy populations and decimate those already living on the precarious edge. In such an environment, surgery, cancer treatment, and other types of critical healthcare could be set back many decades, and even a simple cut could become life-threatening. While I still don't foresee a bacterial strain causing a pandemic, this underscores the reality that while we think about or deal with pandemic emergencies, all of our other healthcare challenges don't go away. We have to face all of them, all the time. Mother Nature is a multitasker.

So, though the Big One will almost assuredly be caused by a virus rather than a bacterium, bacterial infections will continue, maybe even rise, and complicate the already crushing burden on healthcare. And when a microbial infection snowballs into a pandemic, stopping the microbes themselves becomes our most urgent task, as well as the most formidable.

MICROBIAL MYSTERIES

Every human and animal disease outbreak starts as a detective story. In our thought experiment, when Dr. Adamu Kimani observes with dismay, "We may be looking at a world about to change," he knows he and his fellow disease detectives are just beginning to unravel the clues, trying to identify what impact the novel corona virus they've just discovered will have and how much of a threat it represents to humans individually and society at large. When we take into account all the effects enumerated previously, the scary part is not knowing what they don't know.

Like police detectives, disease detectives usually investigate several suspects at once, not knowing which, if any, will turn out to be killers. Keeping track of potential infectious disease outbreaks around the world, we epidemiologists often feel as though we're playing whack-a-mole, always on the lookout for the next threat to pop up.

As you can see from our opening scenario, stopping an airborne, highly transmissible viral pathogen from reaching pandemic potential is extremely difficult, if not downright impossible. How many individual carriers traveled to how many different places before healthcare workers and physicians even realized something was happening? We talk of certain infected people as "superspreaders," people who, for reasons we don't fully understand, transmit certain viruses far more readily than others do. But as we hinted at in our scenario, the greatest facilitator in our modern world is not a human but rather a machine: the commercial airliner. Given that we can't halt an airborne viral pandemic in its tracks, we must be prepared to go after it with every weapon in our public health arsenal.

It is easy to imagine that at the same time as the new coronavirus of our thought experiment was emerging in Africa, epidemiologists would be watching an emerging influenza strain with pandemic potential coming out of Asia. The strain in question could be H5N1 or H2N2 (influenza is characterized and classified by the particular hemagglutinin [H] and neuraminidase [N] proteins on its surface) or any other H/N permutation. As we have seen so often in the past, a new strain of flu can arise and make itself known when and where we least expect it—as in 2009, when an H1N1 flu, a descendant of the 1918 strain, emerged in Mexico, migrated north into the United States, and quickly spread around the world. Though the 2009 outbreak qualified as a pandemic due to its wide international spread, that strain mercifully never reached the scale of the 1918 event. Still, a WHO-CDC study estimated that it caused at least a quarter of a million excess deaths worldwide—that is, 250,000 more people died that year than would otherwise have been expected.

Prior to 2020, we focused on influenza as the most likely candidate for the Big One. Coronaviruses are quite common and were not much of a problem, accounting for the majority of common colds. But SARS, MERS, and then SARS-CoV-2 made clear that we needed to include coronaviruses on the short list of virus families to watch— hence our use of a novel coronavirus in our thought experiment.

Just as we didn't imagine a coronavirus causing a pandemic prior to the emergence of Covid, we can't ignore the possibility that a new or previously unknown virus with pandemic potential could emerge. Some in public health refer to this theoretical threat as "Virus X." But since it remains theoretical, for the sake of realism we chose to go with a virus family that has already proven itself as a global threat.

Among coronaviruses, SARS was initially thought to be a severe influenza strain when it first appeared in Foshan, a city in the Guangdong Province of China, in November 2002. About 525 miles from Wuhan, Foshan was home to animal markets like those in the crosshairs of the initial Covid-19 outbreak. Many of the earliest cases were among food handlers and people living near such markets, with healthcare workers affected next. We described the story of the SARS 2002–3 epidemic, and my work in it, in our previous book, *Deadliest Enemy*. Suffice to say here that, in the end, 916 people died,[2] and in terms of economic impact, the World Bank estimated the outbreaks caused some $54 billion in losses.[3] And that wasn't even a pandemic.

So, what did we learn from the SARS 2002–3 epidemic? First, it underscored the likelihood of future diseases arising from conditions where humans and animals are densely packed and interacting in close proximity, which is realistically not something we will be able to change anytime soon but something we need to vigilantly monitor. Second, the high infection rate among workers at the hospitals where sick people were treated, and among cases that could be traced back to one hotel in Hong Kong,[4] drove home the importance of following the Precautionary Principle, the maxim noted by Dr. Erin Thomas in our opening scenario: *Until you have evidence to the contrary, act as though every new infectious agent is easily spread through the air.* In other words, with something as dangerous as a novel agent with pandemic potential, react as if you are dealing with the worst-case scenario, then scale back if and when evidence suggests less severe measures will suffice. We cannot stress enough the critical importance of this simple concept, which we'll discuss in more detail in Chapter Two.

From the initial SARS epidemic, we saw how the interconnectedness of daily life, as well as the impact of worldwide air travel and its necessity for business, could enhance the transmissibility of any microbe. We also got a glimpse of how devastating the effects of an unknown respiratory-transmitted virus that in some people was highly infectious could be for healthcare systems and both local and global economies. Yet at the time I couldn't help but think that as bad as it had been, it could have been so much worse. I had only hoped we could leverage what we'd learned to prepare for and react more nimbly to the next threat.

It was that thought and my experience with SARS that led me to predict at a 2015 conference at the Institute of Medicine (since renamed the National Academy of Medicine) in Washington, DC, that MERS would show up somewhere outside the Middle East as soon as an unknowing superspreading individual got on a plane and landed in some large and crowded city, as our fictional Indonesian businessman Taramin Wenda did, who felt fine but had ample opportunity to be exposed to the novel coronavirus and transmit it to others. Sure enough, less than two months after that 2015 meeting in Washington, a sixty-eight-year-old man returned to his native South Korea and became ill after having been in four Middle Eastern countries. It took nine days for doctors to come up with a diagnosis, during which time he had been in four healthcare facilities and infected more than twenty people, including personnel at two of those facilities: St. Mary's Hospital in Pyeongtaek and Samsung Medical Center in Seoul. Ultimately, Samsung Medical Center was forced to close to new patients for five weeks, 16,000 people were quarantined, nearly 3,000 schools were temporarily shut down, public events were postponed, and more than 100,000 trips to South Korea were canceled. The outbreak so affected the economic health of the country that the Bank of Korea slashed interest rates to a record low and worried that it would all lead to a depression. The final toll was 36 deaths out of 186 confirmed cases, demonstrating what even a moderate outbreak of a serious infectious disease can do to a modern, sophisticated society.[5]

I'm well acquainted with that outbreak because I made a trip to Samsung Medical Center a few months later at the request of the hospital's president, Dr. Jae-Hoon Song, to consult with him and his team on how to prevent or manage future SARS and MERS crises. The realization that a single traveler could show up in an emergency room and spread an infection if it were not quickly identified and the patient were isolated led us to the conclusion that constant vigilance and early detection of such cases was crucial. Most MERS patients will transmit to only a few people. But some, like the case at Samsung, were superspreaders. Although we still do not understand what caused some infected individuals to be such efficient viral transmitters, it was clear then how important it was to try to quickly identify and isolate all cases. And remember, this same situation could be playing out simultaneously in hospitals all over.

Among the key criteria we use in determining how a microbial agent might spread are ease of transmission and time until onset of infectiousness. Just as real estate professionals talk about "location, location, location," those of us in public health focus on "transmission, transmission, transmission."

Airborne transmission, which is primarily aerosol-related, means the microbe can transmit simply through an infected person exhaling and someone in relative proximity inhaling the same microscopic particles. As we've noted, such a virus with wings can be considered a worst-case scenario, or one most likely to develop into a pandemic. This is the primary way influenza and Covid-19 spread.

Fecal/oral transmission occurs when an infectious agent is contained in an infected individual's feces and, often due to poor hand hygiene, fecal material is ingested orally. Polio is transmitted this way, which is why swimming pools were often closed during outbreaks prior to vaccine availability, since microbes from one swimmer's feces could migrate to another swimmer's mouth.

Close bodily fluid contact is how HIV and Ebola are transmitted.

A break in the skin, such as a bite, can transmit rabies. A scratch or cut also can transmit anaerobic microbes (those that don't need oxygen to survive), such as tetanus bacteria.

Blood transfusions containing harmful microbes are another way of transmitting disease. Before we knew much about HIV infection, for example, too many people, including tennis legend Arthur Ashe, developed AIDS from infected blood transfused during surgery.

From a pandemic perspective, the method of transmission we most worry about and the one that is most challenging to deal with is the airborne/respiratory route of a virus. The other means are easier to control, regardless of the severity of the disease caused by the microbe.

Think about rabies, for example, which has a fatality rate of practically 100 percent if not treated and which causes 60,000 deaths a year worldwide.[6] The fatality rate and horrible disease and death the virus causes would be even greater if it were more easily transmitted.

Fortunately, rabies is not characterized by respiratory transmission. You need to be "injected," usually through a bite, with the bodily fluids or secretions of an infected person or, more typically, an animal, for the virus to spread (although there have been rare instances of respiratory spread in laboratories). We know to avoid dogs or wild animals acting strangely, and to seek immediate medical attention if bitten. Now, imagine that instead of being transmitted via blood or saliva, rabies traveled through the air, and you could become infected just by breathing in virus exhaled by an infected animal you never even saw. That would be a game changer. In fact, I can think of no science fiction setup more terrifying. But it would also be a game changer for the virus itself, because if it killed off all potential hosts that efficiently, eventually there would be no animals or humans left for it to infect, and the virus would die out. Evolutionarily, microbes have to walk a careful line, being infectious enough to keep reproducing but not so deadly that they leave no hosts left to infect.

Unlike rabies, SARS and MERS both transmit through the respiratory route. Thankfully, though, most people infected with either of those coronaviruses have only limited viral infection in their upper airways, meaning that, excluding those outliers who are tremendously efficient at spreading virus, the amount of virus emitted from their respiratory tract is limited. And since, according to the previous outbreaks of those viruses, infected individuals don't generally transmit until they are visibly sick—usually the fourth or fifth day of showing symptoms[7]—there is time to isolate and treat them, thereby holding transmission in check.

So, when I learned the virus that caused Covid-19 was a coronavirus, I was optimistic that it could be contained by using the strategies employed to halt the spread of SARS and MERS and implementing the recommendations we made for Samsung Medical Center. I abandoned that belief within a few days as I closely followed the alerts coming out of places like Hong Kong, and as I spent time on the phone, often several times a day, with researchers in China and media contacts in Wuhan. They all told me not only that transmission was ongoing but also that people were becoming infected who'd had no known contact with anyone who was ill.

We thus found ourselves scrambling with SARS-CoV-2 because its transmission was dynamic, much more so than that of SARS or MERS, and it was often occurring *even before the onset of symptoms.* That was one of the most frightening "light bulb" moments of my entire career. When I realized the novel coronavirus was being spread throughout the world by people who often didn't seem sick and/or didn't know they were carriers, I felt as though we were leaning over the railing of a hundred-story building, with nothing to catch us if we fell. We didn't have a playbook for a coronavirus like that. A stealth infector, transmitting through the air and into the nasal passages and throat — that was the stuff of an epidemiologist's nightmares. And it's something we need to factor into our planning for future pandemics.

While some victims of Covid-19 became extremely ill and died, others never felt a single indication they were infected and infectious. It remains one of the great mysteries of SARS-CoV-2, and I fear that the Big One will spread in a similarly stealthy fashion, which is why I've chosen this attribute for the novel coronavirus of our thought experiment.

To those who argue that pandemics are "optional," or that we can *prevent* them, I ask, How can you stop something you don't know is there? We can and must *prepare* for pandemics, and we can and must develop strategies to mitigate their effects, but when presented with a highly infectious virus with wings that can be spread by pre- and/or asymptomatic individuals, we are unlikely to stop it before it reaches the pandemic level. Think of how China tried to achieve zero-Covid by essentially locking down everybody and everything, everywhere — an approach few countries would be able to impose on their citizens even if it could be shown to work — and yet they were unsuccessful. As soon as they lifted restrictions, the virus swept the nation. And keep in mind, this was an emerging pandemic that was already recognized. How do you impose such preventive measures on something you haven't yet detected or identified?

The challenges ahead are clear, and they represent the greatest threat to national security that we, and most other nations, will ever face. Meeting those challenges will require public understanding, scientific and intellectual resourcefulness, significant financial resources, political will, international cooperation, and social comity. Admittedly, that's a pretty tall order, but then again, the microbes don't care.

TAKEAWAYS

1. It will be impossible to stop a highly infectious respiratory virus—one with "wings"—from causing a pandemic. We can only limit its global impact and mitigate its effects, and even then, only if we are prepared. Nature will have the upper hand, and the success of our response will be directly tied to the extent of our preparation.

2. Prior to Covid-19 we believed that influenza viruses were the only infectious agents capable of causing a respiratory disease pandemic. Now we realize that we must prepare for the emergence of both novel influenza and coronaviruses of pandemic potential, while also keeping an eye out for other emerging virus types.

3. When the next pandemic hits, whatever preparations we have (or haven't) invested in, the safest course of action for everyone—frontline healthcare workers, first responders, and everyone else—is to follow the Precautionary Principle advised by Dr. Thomas in our thought experiment: *Until you have evidence to the contrary, act as though every new infectious agent is easily spread through the air. It could save your life.*

CHAPTER TWO

THE AIR WE BREATHE

*We must remain devoted to the canons of science that are so
much part of the practice of medicine and the practice of the
allied arts. To do otherwise would be to build public policy on
quicksand.*

— David Axelrod, MD, former commissioner,
New York State Department of Health

[Pandemic Day 18]

After putting out a notice of the new coronavirus, WHO sponsored an international video call of more than a hundred participants, including representatives from the EA-RCC, AFRO, various WHO regional offices, the ECDC, and the US CDC, as well as some of the leading coronavirus experts from the SARS-CoV-2 pandemic. It was exactly ten days since baby Hani Axlam Omar died at the hospital in Hagadera.

Chaired by Jeremy Davies, the executive director of WHO's Health Emergencies Programme, and closed to the public and media, the virtual meeting centered on what was known about the novel virus, where it had been detected, what kinds of testing were currently available, recommendations for the public, and the challenges that would be faced as each new region or country experienced the pathogen. WHO had learned of at least eighty-seven "suspect cases" in at least nine different countries. Forty-two of these patients had died, most in east central Africa, though the dead also included a twenty-six-year-old French NGO volunteer who had been working for several weeks at the

Hagadera Refugee Camp in Kenya. Twenty percent of the deaths were children under five.

Davies first called on the experts from Kenya. Dr. Joseph Ndembi presented epidemiological data on the cases, and Dr. Adamu Kimani reported on lab findings. As the first physician to recognize the virus, and having treated numerous patients since then, Dr. Daniel Onyango also had been asked by Davies to present to the large panel of experts. He was clearly exhausted, and his voice broke as he mentioned the children he could not save. Later, participants would remember his speech frequently punctuated by a hacking cough, and the physical distress as he drew breath. It would turn out to be the last time any of them would hear his voice.

Following reports from the local experts, Davies opened up discussion to representatives from other countries with suspected cases. Several participants noted the challenge of working from just the general case definition with little else to go on. WHO, in conjunction with in-country experts, planned to develop a more comprehensive case definition within forty-eight hours, which would be communicated to public health agencies worldwide. Even with that, there was general agreement that tests needed to be developed and widely available immediately. Laboratorians from Europe and the United States indicated they were already working on PCR assays. The team from Germany reminded everyone that although they had a test for Covid ready in seven days, securing enough reagent and testing supplies to make it readily available worldwide was just as much of a struggle now as it had been in the early days of the Covid-19 pandemic.

WHO conveners attempted to address, but couldn't answer, the cluster of questions: How fast can we get a test, and for whom? How about a vaccine? How will we ensure tests and vaccines are distributed equitably and not first to high-income countries, particularly since so many of the cases are in Africa? How can we protect healthcare workers? Do we know how this is spread? Is it airborne? Where did it come from? Did it escape from a lab? What do we want political leaders to do? What do we tell the public? Will previous SARS-CoV-2 infection or vaccination provide any protection? Do we need to lock down affected areas? What about closing borders?

Representatives from the United States, Europe, and Asian nations addressed one of the questions, confirming that they'd already begun working on an mRNA vaccine for this virus, although no one could commit to a time frame for when it would be ready for testing, let alone for public use.

As Dr. Tamara Goldfield, the newly appointed CDC director, listened to her colleagues from around the world, she knew the mortality rate was likely not as high as the nearly 50 percent the numbers so far would suggest because milder symptoms would not have been included in the case definition, and many cases that recovered on their own probably hadn't been reported. Also, the baseline health condition of many in east central Africa and their limited access to healthcare would sadly put them at greater risk of a more serious outcome. Still, if the mortality rate was just one-fifth of what it seemed, the implications were alarming. Even more disconcerting, however, was Goldfield's sense of déjà vu while listening to the confusion and seeing the lack of medical and scientific leadership. The local experts from Nairobi were on top of things, as best as they could be, but it was already clear this was not going to be run as a united international effort. As with Covid, individual countries were already gearing up to handle things on their own.

Eventually, Davies reined in the questions, saying that more information would be provided as soon as it became available. In perhaps the only clear action to come from the meeting, WHO suggested that the tentative designation for this novel virus should be, simply, SARS-CoV-3. The WHO participants signed off to prepare for another conference, promising to reconvene the group the following day. The next conference, later that afternoon, would be with the major world media.

―――――――

In Kenya, it had taken the epidemiology field staff two days to formulate a strategy to test individuals and animals back in the earliest environments where the contact-tracing interviews suggested the virus might have originated. Based on information received from contacts in Eastleigh and Hagadera, additional teams were sent into Somalia—to Burgabo, to the area bordering Lag Badana National Park, and into the park itself. Though private farming and hunting in the park were against the law, because of frequent, extended drought conditions, people were driven there for survival, and the area was rich in animals that could harbor a coronavirus.

Those sent to the park wore personal protective equipment (PPE). They were specially trained to take samples from monkeys, bats, hyenas, gazelles, foxes, and any other animals that could be a natural reservoir or vector for the novel coronavirus. The team in Burgabo and at the border had an easier time because most of the animals they sought for testing were in the control of humans. They took samples primarily from camels and donkeys, but also from goats and other domesticated animals. Both camels and donkeys were known to sneeze and cough, and camels had the nasty habit of spitting.

The team also sought out the area's community health worker, Jamilah Shamshi, who told them about Warsame Amir Osman's family and the many people she'd seen with the new illness since Warsame and Cabdi died more than ten days ago. They realized she'd likely been the first person to examine anyone presenting with the new disease. Distressingly, by the time they found her, she was also coughing and wheezing. She said she was simply worn out from having tended to many more sick people than usual, several of whom had died. She was reluctant to leave her responsibilities, but the team convinced her to go to the hospital, telling her that if she was sick and gave to it anyone else, she would be doing no one any good.

———

A few hours after the video conference, WHO director-general Dr. Kolawole Adebayo joined Dr. Davies and a smaller group of WHO scientific experts on an international media call, covered live on CNN International and the BBC. Although only WHO personnel would conduct the briefing, representatives of other key institutions had been invited on, including Dr. Jonathan Ballard, director of the CDC's National Center for Emerging and Zoonotic Infectious Diseases. He listened as members of the media asked questions that, while reasonable to the situation, were virtually unanswerable until there were more data on the new virus.

Jamie St. Clair of the Associated Press asked, "Are you recommending that nations restrict flights into their international airports to limit transmission?"

"No," Davies said. "Data from studies of previous infectious disease outbreaks, including SARS-CoV-2, do not show that border closings stop, or significantly slow, disease spread."

Tom Banks of Reuters asked, "Does that mean local lockdowns are needed to protect the public?"

Dr. Adebayo frowned as he answered, "First of all, it's not clear what we're talking about when we say 'lockdown.' Whatever is to be done will be left up to each country to decide."

Banks quickly followed up. "But you just said the virus is everywhere. What kind of surveillance and testing is being done to make sure we know specifically where it is?"

How can we even know, *Ballard thought to himself,* with surveillance as limited as it is? *There were no specific PCR or rapid tests available for the novel coronavirus, except experimental ones in several research or public health labs. And even if there were, global reagents, swabs, and everything else would be in limited supply.*

Dr. Adebayo attempted to respond to Banks. "We are still in the early stages of identifying cases. As we know more, we will provide additional information."

"Nadia Gonzales, CNN: Do we know if people have any protection against the new coronavirus from either Covid-19 vaccination or previous infection? And will we be able to quickly modify those vaccines to fight the new virus?"

Davies jumped in. "Those are good questions, and they are things we will certainly be looking into as we learn more about the virus."

Kate Brenner of The Guardian asked, "Is this virus spread through airborne transmission?"

"That may be a component," Dr. Adebayo replied. "We are investigating that now."

"I believe this was a major point of contention with Covid, and it took WHO quite a while to acknowledge that it was airborne," Brenner followed up. "Are public health organizations going to be on the same page this time around so people won't be as confused or divided?"

"We will provide the best guidance as information becomes available," Dr. Adebayo asserted before ending the call.

————————

The next day, a briefing from the White House press room provided the US government's first official communication to the American public about the new virus. Press secretary Deann Morgan introduced CDC director Goldfield, backed on the podium by Dr. Paul Richman, head of the Office of Pandemic

Preparedness and Response Policy (OPPR); Dr. Brian Caldwell, director of the National Institutes of Health; Dr. Caitlin Malone, the Department of Health and Human Services assistant secretary for preparedness and response, commonly referred to as ASPR; and Alejandro Borges, director of the White House Office of Science and Technology Policy.

In her remarks, Dr. Goldfield stated that there had been five presumed cases of the new coronavirus in the United States so far. One person had died, and the other four were in critical condition. All were known to have traveled from eastern Kenya or southwestern Somalia, or had been in close contact with someone who had been in those areas, within five days of the onset of symptoms. She said she suspected there were other cases in the country that were asymptomatic, presymptomatic, or experiencing only mild illness and therefore had not sought healthcare but could still be infectious. Goldfield explained that the CDC was working closely with state public health labs to develop a PCR test, and that OPPR, together with the Food and Drug Administration, was coordinating with the private sector to develop additional tests that would be available to hospitals and local doctors' offices as quickly as possible. In the meantime, state public health labs would be working to rule in or out suspected cases. In closing, she urged anyone who met the case definition and criteria — anyone who had the respiratory symptoms, fever, fatigue, muscle ache, and, in some cases, diarrhea; and who had traveled in or been exposed to someone who'd recently been in Kenya or Somalia — to report their illness to their local or state health department, even if they weren't sick enough to require hospitalization. She turned the mic back to Morgan, who opened the floor to the gathered reporters.

She faced a sea of raised hands and urgent questions:

"We're getting reports of hospitals saying they think they have cases, but since they can't test, they can't be sure. What are you advising them to do, exactly? Are the 'worried well' likely to overwhelm hospitals and emergency rooms?"

"Are you closing the borders to flights from Africa? Will other travel be restricted?"

"What should people do to protect themselves? Do we need to wear masks? Are there enough masks in the Strategic National Stockpile for healthcare workers and first responders? Should people make their own masks?"

"Will schools and/or businesses be shut down this time?"

"Will the president be appointing a SARS-CoV-3 czar, like Ron Klain during Ebola, or White House Covid coordinators Jeff Zients and Dr. Ashish Jha?"

Morgan handled the questions as best she could, but there were few definitive answers to be had. That evening, commentators on the cable news programs showed excerpts of the briefing and noted it was hard to tell how concerned anyone should be about this new virus. The country was still deeply divided over Covid-19. Some felt dread and fear, remembering the early days of that pandemic. Others, who likely got their news from different sources, were unconcerned and focused on going about their lives, wary of restrictions and perceived government overreach. Meanwhile, though, the news outlets that many of them watched presented several theories about where the virus came from, how it "snuck" into the country, and what the administration's agenda was in promoting news about it.

To the trained observers of infectious diseases, however, it was clear the world was once again on fire.

DR. JOHN SNOW WAS a member of the Royal College of Surgeons and a pioneer in the safe administration of anesthesia. In fact, he administered chloroform to Queen Victoria during the birth of her last two children.[1] Every few years, London would be hit with a cholera outbreak that would sicken, kill, and spread fear throughout the metropolitan area, and the outbreak of 1854 was the worst in memory. One of the most severely hit areas was Soho, which had inadequate sanitation and sewer facilities. The mortality rate there from the epidemic exceeded 12 percent.[2] Dr. Snow wanted to know why.

The strategy he employed against the cholera outbreak remains the basis of public health epidemiology today: observation. Tracking house by house, Snow recorded the cases and noticed the largest cluster concentrated along Broad Street. Then came his critical realization: Nearly all the victims had taken water from the public pump on Broad Street.

Snow went before the local governing body, the Board of Guardians of St. James Parish, presented his statistical analysis, and strongly argued his case. After more than a little resistance to the idea, the board agreed to remove the handle from the Broad Street pump, and the epidemic stopped. As it happened, the number of cases was already on the wane by then, as residents had begun heeding Snow's warnings or had drawn their own conclusions about avoiding the Broad Street pump.

The point of this story is that when we're faced with a disease outbreak, epidemic, or pandemic, the first challenge is to mitigate its effects, and we may have to do so with imperfect knowledge about the causative agent and before we have effective vaccines or drugs. We try to stop the spread the way John Snow did, through what we in public health refer to as nonpharmaceutical interventions (NPI). In the case of SARS-CoV-2, NPI mitigation before the availability of a vaccine meant trying to halt transmission by closing down the places where people were likely to spread the infection, such as offices, restaurants, theaters, and schools, and asking people to isolate at home as much as possible and take precautions when going out to places they couldn't avoid, like grocery stores and pharmacies. These precautions would include wearing effective respiratory protection and practicing what became known as "social distancing," or what I call "physical distancing." If we don't have effective antivirals and vaccines when the Big One hits, much of our initial response will be similar to John Snow's. In the following chapters, we will explore preparation for the inevitable pandemics to come and see how each of our weapons and tools would be most successfully deployed.

The first obligation of anyone in a position of responsibility involves understanding, accepting, and acting on the realities of the challenge they're facing.

Five years before the 1854 London cholera outbreak, Dr. Snow published a paper titled "On the Mode and Communication of Cholera." Even though acceptance of the germ theory of communicable disease was some way off, and the *Vibrio cholerae* bacterium had yet to be discovered, Snow was convinced that the prevailing belief in the medical community was wrong: The disease was not caused by "miasma," or bad air, thought to be emanating from nocturnal vapors in the Thames. He was waiting for actual evidence that could be acted on.

The medical establishment's reluctance to accept Dr. Snow's unpopular idea is not unusual. Whether it was the Hungarian physician Ignaz Semmelweis trying to convince hospital colleagues that simple handwashing by doctors could dramatically reduce maternal mortality by preventing postpartum infection, or French chemist and microbiologist Louis Pasteur building on Semmelweis's observations, which were furthered by English physician Joseph Lister and German physician Robert Koch in substituting germ theory for miasma, doctors, it seems, have often been among the toughest lot to get to change their minds regarding their own orthodoxies. Were it not for these four visionaries, it might have taken many years for medicine to have advanced as it did.

Medical science is often a matter of nuance and degree, and this remains difficult for many people to accept, as so much of our collective Covid-19 experience proved. Pasteur wasn't claiming that one couldn't get sick from the air; he was merely—and importantly—claiming that it wasn't miasma, or the air itself, causing cholera and other diseases. He asserted, and ultimately proved, that something *in* the air—a microorganism—was the culprit, similarly to how Snow surmised that something in the local water supply was responsible for the sickness and death. Dr. Semmelweis died tragically and, ironically, of sepsis in a Viennese mental institution only a few decades before Pasteur proved the validity of his research and advocacy.

I wish we could say that these lessons are confined to the earlier and less enlightened annals of medicine, but the same reluctance to accept new ideas or observations about the means of infectious disease transmission is still with us. The modern equivalent of the miasma theory also involves the air, and it confounds me that so many in public health and medical science are still so resistant to embrace clear and compelling evidence of airborne transmission.

As our fictional reporter Kate Brenner noted in the last question in our scenario's WHO press conference, recognizing the specifics of transmission was one of the greatest points of contention that arose during the Covid pandemic, as it has been in our evaluation of other infectious diseases. And of all the aspects of our Covid response that went wrong, I maintain that the most consequential failure of the public health and medical establishments was not understanding, accepting, and promoting what we needed to do to effectively address the critical role of airborne transmission of SARS-CoV-2, meaning via both droplets and aerosols. (While many people think of spray cans of deodorant, air fresheners, and similar products when they hear the word "aerosol," in science and public health, an aerosol is any particle, regardless of size, that is suspended in the air.) Smoke, fumes, rainy mist, even dust from a volcano eruption can be aerosolized. Smoke from a large forest fire can travel hundreds of miles through the air, as we in the northern United States saw in the summer of 2023 when our air quality was affected by wildfires in Canada.

Those of us who've studied the evidence recognize that most of the transmission was aerosol-related, though I have colleagues in medicine and public health who still disagree. To me, their conviction that SARS-CoV-2 transmission is only or primarily the result of droplet exposure within six feet of an infected person is akin to the denial by those who opposed Snow, Semmelweis, Pasteur, and Koch.

A DAUNTING CHALLENGE

Respiratory transmission of dangerous microbes is not only among the most crucial considerations in all of epidemiology but also one of the most daunting. You can diet in an effort to lose weight. You can give up smoking to improve your lung function. You can exercise to strengthen your heart and enhance muscle tone. But you can't stop breathing. And this means that when there's a potentially deadly threat lurking invisibly in the air, each breath we take can be a pathway into our bodies—a route that is accessible to both droplets and aerosols.

I am not an aerosol science expert, but I work with some of the best in the field, a highly technical discipline. Aerobiology is the study of biologic (i.e., not chemical) particles such as bacteria, viruses, fungi, and pollen. Aerobiologists and their colleagues in industrial hygiene study how infectious agents spread in droplets and aerosols. They conduct sophisticated air testing in laboratories to reach a detailed understanding of how respiratory protection devices work. Now, the distinction between small droplets in the air and floating aerosol particles might seem subtle or insignificant, but to those who specialize in aerobiology, and to epidemiologists like me, it is profound, with tremendous implications for how we approach disease prevention. So, what's the difference, and why is it critical to pandemic preparedness going forward?

Droplets are tiny globs of liquid that come out of your nose or mouth when, say, you sneeze or cough. They can be expelled when you sing or shout, as anyone who's been in the front row at a play or concert may have observed, watching performers' spit project, along with their words or song, illuminated in the bright stage lights. As small and generally unnoticeable as these particles may be, they're heavy enough to fall to the ground by force of gravity. Droplets travel short distances or sink to the nearest surface. This is where the guidance for maintaining six feet of social distancing came from during the Covid pandemic. Many who argued that Covid was spread through droplets figured that since the infected droplets would have to be propelled into mucous membranes to take hold, keeping six feet away from other people would make it unlikely you could catch the disease that way.

Aerosol particles come out of your nose or mouth as droplets do, but if you want to understand an aerosol, think about how you can smell perfume across a wide mall hallway from the stores that sell it, or the last time you saw dust particles streaming in the sunlight coming through a window in your house.

If I'm in a room speaking, within minutes, small particles expelled from my mouth and nose will be floating in the air, even though no one may see or feel them. If you're in that room, you're going to inhale my particles and exhale particles of your own—what I call "swapping air." Droplets come largely from coughing or sneezing, and the droplet hits you in your nose, eyes, or mouth, like an incoming projectile. Compare these droplets to the free-floating aerosol particles circulating from that same cough, sneeze, or even just breathing. The aerosols are present in that same six-foot "social distance" zone as the droplets are, but aerosols are also potentially present even *yards* away. You can see how the transmission of a respiratory pathogen via an aerosol versus a droplet is a game changer in terms of the ease with which a virus can spread. And if the people who are infectious aren't showing any symptoms yet, and don't even know they are spreading it, as was often the case with SARS-CoV-2 and with the new coronavirus of our thought experiment, that ratchets up the risk another significant notch, with no corresponding indication of the threat.

Outside, air currents move and spread the viral particles quickly and far enough apart that the danger of infection decreases, though it certainly isn't nonexistent. But inside, even in a large room, the concentration of viral particles means that we "swap air"—and whatever is suspended in that air—much more efficiently in enclosed spaces.

I understood the potential of airborne virus transmission when I led an investigation of a measles outbreak at a Special Olympics event at Minnesota's Hubert H. Humphrey Metrodome in 1991. The infection was traced back to a twelve-year-old track-and-field athlete from Argentina who was in the early, highly infectious stage of illness when he stood near home plate for several hours during the opening ceremonies. In addition to the event officials, competitors, and support staff who subsequently came down with measles after their exposure to the young man, two cases were later reported in Minnesota residents who did not know each other, had no direct contact with the infected athlete, and had attended only the opening-night festivities up in the stands. In terms of exposure, the only thing they had in common was their location that night: They sat in the same upper-deck section *more than four hundred feet* from where the infected athlete stood and exhaled virus for several hours. Air-circulation patterns at the enclosed Metrodome had allowed aerosolized measles particles to make their way from the ground level all the way up to them, skipping up and over all the rows in between.[3]

The same virus, producing the same symptoms of respiratory infection but instead spread by larger droplets, would be a good deal easier to control and potentially keep from evolving into a pandemic. Most people are now familiar with measures to prevent droplet transmission because we tried them in an attempt to stop SARS-CoV-2 spread. Think back to the early days of Covid, when banks, store checkout lanes, front desks at apartment buildings, etc., put up glass, plastic, or plexiglass barriers to keep people separated from one another. That may have done some good in protecting folks from sneezes and coughs, but unless those partitions provided a complete and unbroken separation from wall to wall and floor to ceiling, with no air mixing between the two sides (making any kind of interaction almost impossible), they weren't accomplishing anything when it came to aerosol particles.

When I saw those kinds of barriers, I'd ask, "If someone was smoking a cigarette on the other side of the partition, would you be able to smell it?"

"Yes, of course" was the usual answer.

"Then you're not protected from aerosol transmission."

This is what we call *hygiene theater*. In July 2020, Derek Thompson, a staff writer for *The Atlantic*, published an article on the subject, criticizing measures being taken during that early stage of the pandemic that provided a false sense of security but did little to reduce the spread of Covid-19.[4] Untold millions, if not billions, were spent on such measures. Imagine if we had, instead, invested some of that money ahead of time into stockpiles of personal protective equipment (PPE)—specifically, N95 respirators—for our healthcare workers, first responders, and essential employees, like those who work at food-processing plants and shipping facilities.

This was one of the greatest failures from the Covid-19 pandemic. It is a lesson we must learn now so we can do better next time. We should have educated everyone that the most effective way to avoid breathing in SARS-CoV-2—an aerosol- *and* droplet-spread virus—was to use an N95 respirator (officially called an N95 filtering facepiece respirator) correctly and consistently. I don't like using the term "mask" to refer to N95s, because lumping them in with all the other masks and face coverings conveys a false sense of equivalence. But since "mask" is in common usage, I've pretty much stopped fighting that battle. Just please know, the only facepiece that provides the kind of protection we need against airborne viruses is an N95 respirator, regardless of what you call it. Using surgical (sometimes called procedure) masks against an aerosol-transmitted virus is the equivalent of putting a screen door on a submarine and expecting it to keep water out. To encourage, let alone mandate, the use of "masks" without both making clear that the only highly effective ones are N95 respirators and explaining how they must be worn for maximum protection was a major mistake.

Still, as appalling as the suffering from the Covid-19 pandemic was, we were lucky in that it had a relatively low mortality rate compared with that of SARS or MERS, or the virus in our thought experiment.

The strategy we need to employ is clear: Start out by treating it as a potentially aerosol-spread virus. This is a lesson we should have learned from hard experience. In 2006, an independent commission formed to investigate the outbreak of SARS in Canada concluded, "When it comes to worker safety...we should not be driven by the scientific dogma of yesterday or even the scientific dogma of today. We should be driven by the precautionary principle that reasonable steps to reduce risk should not await scientific certainty."[5] It is always easier and less costly on many levels to back off from that position later if you learn it is safe to do so than to play catch-up.

AN EXPONENTIAL GROWTH CURVE

When we described Dr. Erin Thomas invoking the Precautionary Principle after her initial examination of Abdirahim Ali Salat in Chapter One's scenario, we were showing an example of someone acting wisely and proactively, seeking to avoid spread of the mystery illness to others on staff and those in their care. What she had no way of knowing, but we, as omniscient narrators and readers did, were the many opportunities SARS-CoV-3 had to spread:

- The virus spread through the Warsame Amir Osman's family to community health worker Jamilah Shamshi and then to those they came in contact with.

- Whether it was already at the camp or not, the virus spread from the Axlam Omar Yussef's family to others at the hospital in Hagadera Refugee Camp, including Dr. Onyango.

- Etienne Navarre, the young French aid worker who eventually died from his illness, brought the virus with him on long flights, including from one of Africa's busiest airports to one of the world's busiest airports.

- Businessman Taramin Wenda had ample potential exposures in Nairobi and Wajir—where he spent time up close with camels, known to be reservoirs of the coronavirus MERS.[6] He never felt symptoms, but his body nonetheless could have harbored virus, enabling it to spread to Indonesia, after he first networked with other international travelers in the Istanbul airport lounge. If he infected any of them, who knows where they spread it?

- College student Curt Ashworth spent nearly twenty-four hours traveling home to Georgia on crowded flights, spending several hours in Frankfurt Airport before continuing on to Hartsfield International in Atlanta. He then stayed overnight in a hotel, waiting in line at the breakfast buffet with numerous unsuspecting travelers.

- Before he made it to the emergency room in Minnesota, Abdirahim Ali Salat rode in a truck three hundred miles from Kismayo to Mogadishu, then sat

in airports and on three flights that took him from Somalia to Qatar, then to Dallas Fort Worth, through US
Customs and Immigration, and finally to Minneapolis–
St. Paul. Think of how many crowded and close-
contact security, boarding, customs, and immigration
lines he'd stood in during his journey.

Now think of all the people who unwittingly participated in this
exponential growth curve, from the countless refugees in the close
quarters of the Hagadera camp to people riding buses with anyone
infected from the camp to Eastleigh, and all the people our air travelers sat near for hours on the long flights or spent time with in airport food courts, bathrooms, and lounge areas. Certainly some of
these people would have taken the virus with them wherever they
were going next, and so it would continue to silently spread. By the
time it came to health authorities' notice, the winged virus had
already taken flight around the world, which is why closing international borders would have made little sense or difference. It doesn't
much seem like this type of pandemic is "optional."

WILLFUL BLINDNESS?

The full story of why and how we were unprepared to deal with airborne transmission of SARS-CoV-2 is sufficiently involved that it
could fill an entire book. At its core, though, are two challenges we
need to get past to be able to meaningfully plan for future microbial
threats. The first is a continued widespread reluctance of many in
public health and related fields, including medical practitioners and
policymakers, to acknowledge that airborne transmission is a real
risk, believing instead that it only rarely occurs. The second is the
powerful role of inertia or resistance in these groups to change their
position, even in the face of compelling scientific data.

Dr. Lisa Brosseau is a highly respected and valued colleague of
mine at the Center for Infectious Disease Research and Policy
(CIDRAP) at the University of Minnesota, as well as a leading

bioaerosol scientist and industrial hygienist who has consulted for major corporations, regulatory agencies, and educational institutions. Early on in the Covid-19 pandemic, I shared with Lisa that my gut instinct—based on my fifty-year career investigating respiratory-transmitted viruses including SARS, MERS, and influenza—told me SARS-CoV-2 was transmitted by aerosol.

My suspicion was right: Months and then years of living with Covid yielded more and more data demonstrating airborne transmission. Both Lisa and I were deeply disappointed that so many in our fields resisted acknowledging it as a main cause of Covid spread—just as many still do with influenza, despite the large body of scientific information showing it to be true.

WHO, too, remained in the same magical-thinking camp. Just as it lagged in declaring the worldwide Covid-19 outbreak a pandemic, it was behind the curve concerning transmission, sticking to the position that SARS-CoV-2 was spread primarily by respiratory droplets and physical contact, and focusing on sanitizing surfaces and the hygiene-theater approach even in the face of clear, solid evidence that the virus was airborne. Early in the pandemic, WHO tweeted "FACT: #COVID19 is NOT airborne,"[7] and it persisted with this message for many months, waiting for what it would accept as conclusive evidence of airborne transmission.

Frustrated by this official, incorrect, position, in July 2020, 239 scientists from thirty-two nations directed an open letter to WHO, principally written by Drs. Lidia Morawska and Donald K. Milton. It was published in the Infectious Diseases Society of America's journal *Clinical Infectious Diseases* and widely requoted, citing the growing body of evidence showing "beyond any reasonable doubt" that SARS-CoV-2 was spread indoors through aerosols.[8]

Still, the mixed and erroneous messaging continued. In December 2022, at a briefing before Independent SAGE, an organization of scientists working on their own to provide advice to the UK public and policymakers concerning Covid-19, WHO special envoy on Covid-19 Dr. David Nabarro said Covid-19 was "primarily, a droplet-borne infection [that] may be airborne in certain circumstances, [but] we still in WHO contexts say it's primarily droplet-borne."[9] His remarks set off a public health firestorm on social media, including tweets characterizing his statement as deeply disappointing at best, sadly negligent at worst.

The failure of WHO to communicate the critical importance of airborne transmission of SARS-CoV-2 beginning in the earliest days of the pandemic was made public by Dr. Soumya Swaminathan, chief scientist for WHO in November 2022. She announced at that time that she was leaving her post at the end of the month. In an interview with *ScienceInsider,* Dr. Swaminathan stated, "This [i.e., SARS-CoV-2] is an airborne virus." She regretted WHO not calling it an airborne virus much earlier, based on the available evidence.[10] Lisa observes, "It wasn't a hard concept for some parts of the scientific community, but it was very difficult for a lot of the medical and infection prevention and control people. They think airborne transmission is rare. They talk about measles and tuberculosis [as aerosol-transmitted microbes], and pretty much everything else, they think, is all droplet transmission." She further notes, "The word 'aerosol' does not even appear in the US influenza pandemic plan."[11]

We see this willful blindness again and again. After the first SARS outbreak in Asia, we had numerous warnings from experts concerning aerosol transmission, including a particularly insightful article in the April 22, 2004, issue of the *New England Journal of Medicine* by Drs. Chad J. Roy and Donald Milton entitled "Airborne Transmission of Communicable Infection — the Elusive Pathway." Chad is professor of microbiology and immunology at Tulane University, and Don is professor of environmental health at the University of Maryland. Both are internationally recognized aerobiologists whom I am proud to call my friends and colleagues. Citing the irony of John Snow's struggle to convince London's general board of health that cholera transmission was waterborne rather than airborne according to the miasma theory, Chad and Don examined the SARS outbreak in Hong Kong in the preceding years and presciently warned that "the peculiarity of the SARS coronavirus...may be a harbinger of unorthodox transmission patterns associated with emerging infectious agents in the modern built environment. It is a clear demonstration of the need for better understanding of aerosol-acquired disease—whether airborne transmission is obligate, preferential, or opportunistic—and for improved vigilance and infection control."[12]

They went on to explain: "What underlies the low repute of airborne transmission today? First, the two diseases whose airborne transmission is most widely acknowledged, measles and tuberculosis, had been historically controlled through vaccination and drug therapy. As a result, the impetus to understand the aerobiology of infectious diseases has faded. Second, contamination of water, surfaces, and large-droplet sprays can be easily detected. It is difficult, however, to detect contaminated air, because infectious aerosols are usually extremely dilute, and it is hard to collect and culture fine particles." In other words, we don't want to think about it, because it's too hard. Unfortunately, we can't just wish it away.

Sadly, measles and tuberculosis can no longer be considered well controlled. Measles is ramping up because not enough people are being vaccinated against this highly infectious disease. Tuberculosis cases and deaths are on the rise in many places due to delayed diagnoses, poor and crowded living conditions, lack of funding and adequate healthcare, and antibiotic resistance to certain strains. The strong forces of inertia and lack of impetus to understand the aerobiology of infectious disease, just as sadly, remain.

So, why is there such intellectual opposition and dissent to the very idea of airborne transmission?

Along with Lisa and others, I think there are various reasons beyond the ones Chad and Don laid out in 2004, and they cover several categories of denial. First, there is the political/economic argument asserting that it isn't feasible to protect everyone with a respirator: It's too expensive, and we don't have enough anyway, so why even bring it up? And long-held convictions are difficult to change. If it is hard enough to get people to wear any kind of mask, how much harder will it be to get them to accept a particular kind of mask that is a little more expensive and somewhat less comfortable?

Unfortunately, right from the beginning of the Covid-19 pandemic, there was a proliferation of problematic studies "proving" that methods to prevent transmission—such as wearing N95 respirators rather than surgical, cloth, or even no mask—made little to no difference. So-called experts with no training in industrial hygiene, aerobiology, respiratory protection, or epidemiology wrote analysis pieces or conducted masking studies that were seriously flawed from a scientific perspective. Non-expert journalists would expound on such studies with no sense of how little they actually knew or understood. Of note, some of the most problematic studies were published by the CDC in their highly respected and relied upon *Morbidity and Mortality Weekly Report* (*MMWR*).

The industrial hygiene and aerobiology experts had already determined the science on this issue. And, for the life of me, I can't understand the anti-science stance that persisted in organizations like WHO and the CDC. People's lives were in jeopardy because of their misinformed position!

A third consideration is psychological. As we noted earlier, you can change your behavior and cut back on bad habits to promote health, but you can't cut back on breathing, so the idea of catching something just by inhaling is terrifying. The fear may be subconscious, but many officials don't want to be the ones to say how dangerous aerosol viral transmission can be—particularly to individuals who are more vulnerable for one reason or another, such as their age or a weakened immune system. We believe public leaders often focus on hygiene theater because at least it seems as if they are addressing the issue.

There may also be an element of pride. If you've been saying publicly and loudly for years that a virus is spread via droplets, it would be embarrassing to walk that back.

Of course, none of these is sufficient reason to ignore scientific truth and place people in danger.

MASKING: THE TRUTH ... OR ... MASKING THE TRUTH

This brings us to one of the most controversial and misunderstood aspects of the Covid pandemic, which I fear will raise significant impediments as we try to plan and prepare for future pandemics: the entire subject of masks and masking. There is a lot of responsibility to go around for the misinformation on this topic, from the media to politicians to the public health community itself. Several years into the Covid-19 pandemic, it was still not clear to most of the public if they should be wearing a mask at all, let alone how different types of masks provide different levels of protection, yet this is the most useful information one could have.

There are a host of issues around the subject, from what type of masks actually offer protection and under what circumstances, to how to make PPE available to everyone who needs it, and how to educate people on when and how to wear such equipment. In the early days of Covid, with medical PPE in short supply, we were told to protect ourselves with some sort of nose and mouth covering but leave the N95 respirators to frontline healthcare workers. Many people resorted to bandanas pulled up over their noses, which made them look like outlaws in old Western movies but hardly did anything to keep them from inhaling viral aerosols. This is not meant to mock people who were sincerely trying to protect themselves but to underscore how challenging it was for people to ferret out good information—something we must do better when the Big One hits.

Early in the Covid pandemic, in February 2020, US surgeon general Dr. Jerome Adams tweeted that masks did not offer any benefit to the average citizen in preventing SARS-CoV-2 transmission.[13] On April 1, 2020, the National Academies of Sciences, Engineering, and Medicine, having established a rapid Covid-19 expert consultation committee, reviewed the possibility of a bioaerosol spread of SARS-CoV-2 and concluded, "While the current SARS-CoV-2 research is limited, the results of available studies are consistent with the aerosolization of virus from normal breathing."[14]

That same month, the CDC published "Recommendation Regarding the Use of Cloth Face Coverings, Especially in Areas of Significant Community-Based Transmission," which stated, "CDC recommends wearing *cloth* face coverings in public settings where other social distancing measures are difficult to maintain (e.g., grocery stores and pharmacies) especially in areas of significant community-based transmission" (italics added).[15]

There was not a single scientific paper or other authority cited to support the idea that cloth masks provided any respiratory protection from airborne transmission. Also on the site was a handout from the CDC's NIOSH, the National Institute for Occupational Safety and Health — the government agency responsible for research and recommendations for the prevention of work-related accidents and illness, recognized as one of the world's leading authorities on respiratory protection — with text stating that a surgical mask "Does NOT provide the wearer with a reliable level of protection from inhaling smaller airborne particles and is not considered respiratory protection." Further, it warned, "leakage occurs around the edge of the mask when the user inhales."[16]

Is it any wonder people were confused?

The word "surgical" sounds authoritative and scientific, as if something surgeons use in operating rooms must be the ultimate standard. In fact, surgical masks were never designed to create an airtight barrier between operating room personnel and the air in the room. They were designed, essentially, to keep surgeons and nurses from sneezing, coughing, or otherwise "dripping" into the patient's open surgical field, and to protect them from blood spatter from the wound. And they might not even do that effectively. Lisa Brosseau notes, "In three randomized controlled studies with and without surgical masks, there was no difference in the rates of wound infection. No difference!"[17]

Given that, I was stunned when I saw Dr. Robert Redfield, director of the CDC during the first Trump administration, on CNBC, holding up a paper surgical mask as he testified on Capitol Hill on September 16, 2020, declaring, "These face masks are the most important, powerful public health tool we have [against the coronavirus]. We have clear scientific evidence they work, and they are our best defense." He added, "I might even go so far as to say that this face mask is more guaranteed to protect me against Covid than when I take a Covid vaccine."[18] I found the lack of clarity on the type of mask he was talking about an alarming and irresponsible example of the kind of incomplete, misleading information that had plagued the CDC and other presumably reliable sources since the beginning of the pandemic.

Does this mean we shouldn't have been wearing masks during Covid or that we shouldn't plan to have them for the next pandemic? Absolutely not! What it means is that we need to get better at defining our recommendations and basing our advice on the best scientific evidence available, and at sharing it in the clearest terms possible.

How a mask protects you from inhaling a virus is all about fit and filtration, and only one kind of mask meets the critical requirements of these two essential features. By this, of course, I mean a respirator— a term that is often confused by the public and media with "ventilator" (a machine normally seen in hospitals and other healthcare facilities that helps you breathe) but that in truth refers to a face covering that could well save your life in the next pandemic.

Tests have shown decisively the difference an N95 respirator makes against no protection or any other type of mask. For example, when an infected person is in the proximity of an uninfected, unprotected person, transmission of a sufficient virus dose can cause infection in fifteen minutes or fewer. One achieves only five to fifteen minutes of additional protection with the use of a typical cloth or surgical mask. In contrast, if an infected and uninfected individual are both using fit-tested N95s, it would take about *twenty-five hours* of exposure for an uninfected person to inhale an infectious dose![19] That's how much difference a certified N95 respirator can make.

The special filtration material in N95s is known as melt blown fiber. It's a synthetic material that is electrostatically charged, which allows it to filter out potential airborne pathogenic particles. In laboratory test conditions, N95 respirators can filter at least 95 percent of particles.[20] The real-world efficiency of an N95 filter is generally higher than what is seen in a laboratory setting. China's KN95, as well as Europe's FFP2, Australia and New Zealand's P2, Japan's DS, and South Korea's 1st Class employ similar filtration material, though the fit may not be as tight or precise.

And the fit of a respirator is just as important as the material for optimal protection. Think of swim goggles. There is no leakage through the plastic lens, but if the seal around the eyes is poor, the goggles quickly fill with water. Cloth face coverings and surgical masks do not fit tightly against the face, so major leakage of air occurs with both inhalation and exhalation. N95 respirators, even when not fit-tested as required for occupation-related protection, can create a good seal if placed firmly and tightly on the face. A mask with some empty space is actually more comfortable to wear. That's why N95s are cup-shaped or otherwise provide more space within while maintaining a seal.

Respirators are commonly used in construction, mining, and general manufacturing, by artists working with materials that may be harmful if inhaled, and by people living in areas affected by wildfires. In fact, typically only about 10 to 20 percent of N95s are sold for healthcare purposes—and that's at the peak of a bad flu season.[21] Before the Covid pandemic, N95s were mostly used to protect workers from inhaling dangerous aerosolized chemicals from paints, solvents, and other substances. Such respirators may have a small valve that lets the wearer's exhalation out efficiently while keeping the bad stuff from getting in. In that context, it doesn't matter that the wearer's breath is getting out without being filtered. With the advent of Covid and use of respirators for nonindustrial applications, it became clear that, even though less comfortable, the hospital type of respirator—without the valve—was the one to protect wearers from the particles around them and protect others from their potentially infectious exhalations.

In the United States, to earn N95 designation, a device must be approved by NIOSH.[22] Though part of the Atlanta-based CDC, NIOSH is headquartered in Washington, DC, with research labs around the country.

In hospital settings, it is recommended that N95s be used for a single shift and then discarded or sanitized, though during the shortage of N95s early in the pandemic, some healthcare systems had workers reuse N95s after storing them for five days to reduce virus on the respirator surface or trapped in the melt blown fiber.[23] We didn't have studies on how effective this technique was. But we found during the pandemic that people who wear them outside healthcare settings—for example, during visits to a grocery store—can use them for several days at least, particularly if they are kept clean.[24]

Admittedly, wearing a respirator for a long period of time can be uncomfortable, even for hospital workers who are used to them. Still, if you want to protect yourself from aerosol pathogens, respirators will work in any public setting where you are concerned for your safety.

On top of the relative ineffectiveness of cloth and surgical masks for filtration, our studies at CIDRAP showed that 25 to 30 percent of people routinely wore their masks below their noses.[25] In my line of work, we call that a chin diaper, and it is useless, like wearing a mask in a restaurant only until it's time to eat. If you give the virus an opportunity, you can't be surprised if it finds its way in. And finally, facial hair and respiratory protection don't mix well. If you're wearing your mask—even an N95—over a beard, it is not providing a perfect seal.

The concept of respirator fit and filtration should be easily understood, particularly by the medical profession. Yet throughout the Covid pandemic, most healthcare workers used surgical masks because they and their institutions were misinformed about the role of airborne virus transmission. In fact, on four occasions when I visited hospitals, I was instructed to remove my N95 respirator and put on a surgical mask because that was what the staff had been told was the standard of care. It reminded me of the pushback against Dr. Semmelweis's warning for surgeons to wash their hands before ministering to women giving birth. As you can imagine, I pushed back strongly against this modern correlative! But if this was what some hospitals were doing, how could we expect the public to get it right?

Early in the Covid-19 pandemic we did have a shortage of N95s for critical use in healthcare. However, by late spring 2021, increased domestic production of N95s and the import of respirators from Asia resulted in an adequate supply to support recommendations for their everyday use.[26] On June 30, 2021, the FDA revoked three

Emergency Use Authorizations that were issued in spring 2020 to combat the respirator shortage.[27] Yet the messaging to the public didn't change. It pains me to think of all the lives that might have been saved if it had.

APPLYING THE PRECAUTIONARY PRINCIPLE

At the beginning of any pandemic, we won't have the scientific certainty we want in time to act. When we aren't sure, the only rational course of action is to plan for and treat it like the worst-case scenario, much as Dr. Erin Thomas does in our thought experiment scenario.

This is precisely what was argued in a report released in March 2022, coauthored by twenty-three experts from a broad spectrum of medicine and public health, including Lisa and me. In the report, entitled *Getting to and Sustaining the Next Normal: A Roadmap for Living with Covid,* we wrote, "The default assumption must be that transmission for all respiratory viruses occurs through small particle aerosol inhalation."[28]

Lisa puts it simply: "In the next pandemic of a respiratory infectious disease, consider aerosol inhalation first! If you assume the worst case and it turns out that it isn't, it's easier to work your way back than to have to mount a quick defense you weren't prepared for. And assume asymptomatic transmission until you can rule it out."[29]

What does that mean in terms of preparing for another pandemic, possibly the Big One?

First is simple acknowledgment of something we learned from Covid but that has been woefully underreported. "Thirty to fifty percent of SARS-CoV-2 transmission occurred in households," Lisa notes. "It wasn't all community transmission, and there has been very little from CDC about this."[30] Depending on the individual home situation, there are measures that can be taken. If there are both school-age children and grandparents in a household, for example, routine precautions should be taken if transmission rates in the community are high. Obviously, this is more feasible in some households than in others. If you have seven or so people representing three generations living in a two-bedroom apartment, say, and the breadwinner is an essential worker who has to be in close proximity to other people while on the job, you're going to have a real vulnerability issue. On the other hand, if those people are all roommates in their twenties and in good health, it's not going to be nearly as much of a challenge. But either way, the next time around, we have a right to expect better guidance from the CDC and other health authorities on what people can do to prevent the spread of illness at home, based on the specifics of the particular pathogen.

To achieve that goal, Lisa says, "We need to develop a respirator for the public. It must be comfortable, easy to fit, have a high filtration level, be highly breathable, and reusable. It must be producible at a reasonable cost and have the manufacturing capacity to produce them in large enough volume."[31] This alone, if it were achieved before the next respiratory-transmitted pandemic—and if the public could be shown the critical value of using such a respirator as standard practice—could be a game changer, even if that next pandemic is the Big One. As of this writing, we are not aware of any organization pursuing this research and development goal.

Large supplies of such respirators would be an important component of the Strategic National Stockpile, an element of the national medical infrastructure for pharmaceuticals, medical supplies, and protective equipment, maintained in facilities throughout the United States and in vendor-managed inventory. The stockpile is an outgrowth of the Cold War civil defense programs of the 1950s and '60s, and even earlier initiatives, but it was already degrading long before the Cold War ended. In 1998, after President Bill Clinton read our friend Richard Preston's realistic pandemic novel, *The Cobra Event,* he pushed for and succeeded in funding and establishing a National Pharmaceutical Stockpile program.[32] During the succeeding administration of President George W. Bush, it was expanded to include protective equipment and renamed the Strategic National Stockpile (SNS).

Each time supplies were withdrawn from the SNS, such as during Hurricane Katrina in 2005 and the H1N1 flu pandemic in 2009,[33] the stockpile didn't regain its previous level of supplies or funding. Clearly, we think the stockpile should be kept well supplied for the obvious contingencies. Stock should be actively managed, tested, and rotated out and into use—replaced by newer stock—at regular intervals so it doesn't become damaged or ineffective. While the fabric component of the respirator is stable, for example, there is vulnerability to the wearer if the rubber bands degrade or the nose clips are not intact. During the Covid pandemic, when effective masks were in short supply, the government authorized emergency use beyond any established expiration date.[34] Rotating out and replacing supply in the stockpile on a regular basis would address that issue.

Additionally, there should be government incentives in the form of guaranteed purchases to produce as many of the critical drugs and items of protective equipment as possible here in the United States and other friendly countries. Otherwise, in a future pandemic, we might find ourselves depending on China and India, two of our biggest offshore suppliers of generic pharmaceuticals, active pharmaceutical ingredients (APIs), and medical devices.[35] I certainly wouldn't want to take that chance.

THE AIR INDOORS

Our next challenge is an order of magnitude larger: What can we do to make indoor air itself safer? Beyond office buildings, retail establishments, schools, etc., realistically, we can't expect families to wear masks continuously inside their homes—and at some point, everyone needs to remove their masks to eat and sleep.

First, we need to realize that it's not rocket science; it's more complicated. There are no easy solutions, because buildings are generally not designed to limit person-to-person transmission of infectious particles. And every building is different in its purpose, use, occupancy level, and construction.

After the devastation of Hurricane Andrew in 1992, Florida transformed its building codes to make residential and commercial structures better able to withstand severe, life-threatening weather. This was not easy, with the results measured in months and years. Likewise, earthquake risks have progressively updated San Francisco's building codes over the past century. Though certainly a longer-term prospect than developing the next generation of respirator masks, something similar to Florida's and California's codes needs to be implemented, based on what we now know and are learning about ventilation and infection control. This will be expensive and will have to happen over decades, with the goal of achieving a permanent and far-reaching impact in mitigating epidemics and pandemics. But before we even undertake such an effort, there is so much more research that must be done to increase our understanding of what it will take to reduce infectivity in a wide range of indoor spaces—from living rooms and classrooms to auditoriums and sports arenas.

This new wave of scientific research will be essential for protecting us during the next pandemic, as well as in our everyday lives. In high-income and many middle-income countries, we expect clean water to come out of our taps, and it is our sincere hope that this will someday be true of the rest of the world. By the same token, we should be able to expect safe air in our homes, schools, and commercial buildings. The challenge is that we don't really know what this means.

Imagine if our recent investment to replace and build anew all aspects of our nation's transportation system—from bridges, tunnels, roadways, and airports to railway and subway stations—were based on incomplete engineering science, and no government agency or professional organization had determined standards to ensure safety and structural integrity. Unfortunately, that is where we are today with indoor air quality and the prevention of respiratory-transmitted pathogens.

The current basis for estimating the risk associated with indoor air transmission of airborne pathogens is known as the Wells-Riley statistical model. It's a highly simplified model of a complex process involving multiple aspects of infectious agent generation, transmission, and removal, developed by engineer William Wells

and physician Richard Riley in 1978, based on measles and tuberculosis investigations.[36] The model describes the probability that a
person susceptible to infection who is sharing a room (and so air)
with an infected person is likely to get infected.

The reason the model isn't sufficient is that it has to contend
with too many variables to be useful in a highly infectious airborne
pandemic. Among the considerations we have to think about are the
following:

1. *Duration.* How long will you and/or others be indoors? The
   longer the time, the more indoor air you inhale with invisible
   airborne particles.
2. *Density.* How many people are occupying the indoor space,
   and how many of them are infected, including symptomatic
   and asymptomatic cases?
3. *Distance.* How far apart are people from one another? The
   closer you are to an infected individual, the more viral particles you are likely to inhale, increasing your risk of infection.
4. *Air mixing.* Even though the ventilation system in a room provides changes of incoming and outgoing air, how well does it
   mix in the room? Are there "dead spots" where air is poorly
   moved by the HVAC system?
5. *Air exchange volume.* How many air changes occur each hour?
6. *Infectious dose.* How much of a particular virus needs to be
   inhaled for an individual to become infected?

Recently, both professional and general publications have proclaimed that we are on the verge of an indoor air quality revolution.
Various sources based on Wells-Riley have concluded that approximately five air changes per hour could be sufficient to reduce risk of
exposure.[37]

How do we know this? The answer is, we don't—at least not with
any degree of scientific certainty. I have spent hours poring over the
studies most often cited. Although there is ample evidence that ventilation helps reduce transmission and provides better indoor air

quality, I am not aware of any reliable data that show that *five* air changes per hour, specifically, significantly reduces the risk of SARS-CoV-2 transmission or that of other respiratory pathogens. Five air exchanges are probably better than one or two, but are ten or twelve better than five in preventing indoor transmission? Frankly, we just don't know.

ASHRAE, formerly the American Society of Heating, Refrigerating and Air-Conditioning Engineers, was founded more than a century ago and conducts research, provides educational resources, and publishes technical standards. It operates in more than 130 countries and has tens of thousands of members. In June 2023, the society approved Standard 241 — Control of Infectious Aerosols. In its press release, it asserted that the new standard "establishes minimum requirements to reduce the risk of disease transmission by exposure to infectious aerosols in new buildings, existing buildings, and major renovations."[38] These requirements would include managing infection risk and determining effective rates of airflow and exchange per hour, effective filtration levels, and the best ways to clean indoor air. The work is ongoing.

That sounds encouraging. But how do you actually implement all this in the millions of buildings worldwide even if you do have the data that it will work? The challenge is that every situation is unique — again, think of those six factors, from how existing indoor spaces are configured to how many people will occupy the space, what they'll be doing (sitting quietly at their desks, not moving around much, or singing and dancing, exhaling a lot of particles and moving the air around), and what level of exposure to a given pathogen constitutes risk. Some measures are certainly better than none to "reduce the risk," but can we say anything beyond that simple statement? And how does all this apply to an emergency situation of trying to mitigate a novel viral pandemic with wings?

Even if we do come up with reliable and actionable data, the challenge to retrofit every building in America, let alone the world, would be overwhelming. In the United States alone there are 5.9 million commercial buildings,[39] with an additional 163,000 industrial[40] and 276,000 military buildings.[41] This includes 437,000 preschool, grade school, high school, and higher education buildings.[42] Add to those, 40 million units in multi-unit residential buildings and 92 million single-family homes.[43]

With the appropriate research, which itself will take years, as well as enlightened and modernized building codes, it is going to be a long time before all building ventilation reaches the level needed to protect inhabitants from the worst effects of microbes. It can begin, certainly, with new construction. For retrofit of existing private homes and public buildings, we will have to prioritize. We don't know, for example, how to retrofit buildings that rely on radiators rather than forced-air systems for heat, as many of our school buildings do. With many building codes starting to take carbon reduction goals and ongoing climate change into consideration, it makes sense to look at these issues together. But the message for our purposes is a sobering one: We're not going to be able to count on "smart" buildings to protect us from viral transmission by the time the Big One comes around. Let's just accept that reality and move on for now.

PLANES, TRAINS, AND BUSES

As we observed earlier, mass transportation — from airplane travel to school buses — offers another infection transmission and control challenge.

First consider school buses. In contrast to adults who ride buses, children often don't stay in their seats on school buses. They sing and shout and often move around to visit with their friends, which can increase the spread of infection. There is also the risk to adults on the bus, including the driver and any aides or chaperones, as well as children with comorbidities or disabilities that make them more vulnerable. However, in many areas and regions, there is no other practical or reliable way for children to get to school.

A driver on a public bus may not have to contend with the chaos of active children, but he or she will have new people continually entering and exiting a fairly closed environment with minimal air circulation, which makes this another potentially high-risk scenario, especially if one is stuck sitting in traffic for hours.

Limited actions can be taken to improve respiratory safety on buses other than N95 masks for the operators, at minimum, and effective ventilation, which will vary from bus to bus. In some situations, opening windows may be impossible, and if occupants are packed together—as in the several-hours-long trip on crowded buses between Eastleigh and Hagadera in our thought experiment, for example—conditions are ripe for the virus to spread.

Commuter trains, subways, and airplanes offer similar potential for viral transmission, particularly before a pathogen has been identified. Picture again Taramin Wenda of our scenario, traveling about without knowing he was infected with a dangerous virus. There are infinite possibilities like this around the world. By the time an unknowing carrier reaches a city center, who knows how many hundreds of people may have been infected? And each one of them becomes a potential carrier in an exponential spread.

Frequent opening of doors at station stops helps somewhat, but active air-handling steps will have more impact. Even before Covid, BART, San Francisco's Bay Area Rapid Transit system, stated that its railcars filtered and exchanged inside air about every seventy seconds,[44] which is impressive. In 2022, Washington, DC's Metro system reported air exchanges in their Metrorail cars every three minutes,[45] which was greater than in most buildings. In the wake of Covid, Metro received a grant from the Federal Transit Administration to evaluate the effectiveness of ultraviolet-C light filters, sometimes referred to as germicidal lamps, to kill bacteria and viruses in the cars' air-handling units. They are also experimenting with the installation of MERV-13 (Minimum Efficiency Reporting Value) air filters that can trap particles as small as viruses.[46] The Environmental Protection Agency doesn't claim that MERV filters alone will fully protect riders from all viral particles, but they are a sensible part of an overall protection program. We hope that Metro's pilot program and those in other large city commuter rail systems will be completed and give us data on what works best before the next pandemic emerges.

WHAT CAN BE DONE NOW

As we await research findings, and funding, for large-scale strategies, there are smaller, room-size measures that have proven useful in cutting down on transmission. HEPA (which stands for "high-efficiency particulate air") and MERV filters were studied during Covid-19. The April 7, 2023, issue of *MMWR* examined various ventilation approaches, concluding that maintaining continuous airflow movement in school buildings was the least expensive means of effective ventilation, noting that it was employed by about half of the school districts surveyed, while portable air cleaners with HEPA filters were reported to be in use less frequently.[47] HEPA filters are similar to those used in N95 respirators, though they are even more efficient at removing infectious particles.[48] The efficiency of MERV filters, used in building ventilation systems, varies based on their rating. MERV filters with lower ratings, which are often primarily intended to remove larger particles like pollen and dust, cannot efficiently remove small viral particles.[49] They can be useful by keeping the air circulating and removing larger particles that can act as carriers for the virus, but we don't know to what extent the filters actually lower viral transmission.

That none of the ventilation strategies was used by more than half of the school districts, the *MMWR* report stated, "underscores the ongoing opportunity to improve indoor air quality among K–12 school buildings in the United States."[50]

Another supplementary tool is ultraviolet (UV) light, which, when trained on viruses or bacteria, can neutralize their ability to replicate by damaging the nucleic acid of cells. It has been shown to be effective in the air, on surfaces, and even in water, though the larger the area you're trying to disinfect, the larger a lighting array you would need. This makes it impractical for large spaces. A 2021 article in the *Journal of Hospital Infection* notes, too, that the farther away you get from the light, the less effective it is. The article also states that advances in UV technology for killing viruses supported further research and development, and possible application in hospitals and other healthcare settings.[51]

A highly infectious virus with wings is, by definition, adept at the one thing it is designed for: transmitting copies of itself from host to host so it can keep reproducing—basically its only mission in life. Depending on your viewpoint, it is either ironic or a simple fact of nature that by creating indoor spaces—in other words, the shelters we need to survive and thrive—we have helped these viruses do just that.

TAKEAWAYS

1. At the outset of any pandemic, authorities and healthcare personnel should apply the Precautionary Principle: Assume and take steps to protect against highly infectious airborne transmission, such as by deploying N95 respirators, until or unless it can be shown that aerosols are not a significant means of spread and transmission. This is something that ordinary citizens can do, regardless of public policy.

2. The public health, clinical medicine, and media establishments persisted during Covid-19 with messaging and reporting on recommendations for respiratory protection that were not science-based. That misinformation continues to this day and could be a substantial impediment to protecting the public during the next highly infectious, airborne virus outbreak.

3. We must translate occupational health science on airborne virus transmission into our recommendations for how members of the general public can effectively protect themselves. These include designing and manufacturing N95-comparable respirators that are reasonably comfortable to wear for extended periods and washable for reuse.

4. Indoor-air exposure to the pandemic virus is a significant method of transmission and infection, but we can expect only limited

risk reduction in this environment in the future given that we don't yet understand the ventilation metrics and standards required to reduce risk in the millions of buildings worldwide that would need modifications. Therefore, we must assume that, for the next pandemic, indoor air quality will not be at a level that would substantially reduce the risk of exposure, and we should plan accordingly.

CHAPTER THREE

MANDATES

If upon examination any such individual is found to be infected, he may be detained for such time and in such manner as may be reasonably necessary.

—42 US Code § 264 (d)(1) (1960)

[Pandemic Month 1]

It was only the second time Dr. Tamara Goldfield had been in the Oval Office. The first had been when the White House science adviser Alejandro "Andro" Borges and chief of staff Gilbert Stern brought her here to meet the president before he announced her appointment as CDC director. She felt an unmistakable, and frankly intimidating, sense of awe standing in this room, knowing the most consequential decisions in the world were made here each day. And today would be no exception, as the president had gathered his top advisers to discuss what to do about the rapidly spreading virus the world had only formally heard about days earlier. She wished she had more concrete details to communicate from yesterday's now-daily WHO briefing, but this early in, you couldn't really expect to know much.

Looking around the room, she took in a formidable group of experts. Dr. Peter Sadlock, director of the NIH's National Institute of Allergy and Infectious Diseases, stood at the foot of the coffee table between the two facing sofas. Stern had pulled one of the visitor's chairs that flanked the president's Resolute desk into position opposite the fireplace, while OPPR director Paul Richman, Health and Human Services secretary Anne Sulbarry,

Homeland Security secretary Ross Weinberger, NIH director Brian Cald-well, and HHS ASPR Caitlin Malone occupied the sofas. Andro and National Security Advisor Kara Marsh stood grimly to the side.

When everyone was assembled, the president rose from his chair behind the desk. All eyes followed the chief executive as he crossed the center of the room and took his seat next to the vice president in one of the two armchairs in front of the marble fireplace. Goldfield started to rise out of respect, but the president motioned her to remain seated.

He began solemnly, "I want to open by noting, as some of you know, great statesman and personal friend and mentor to many of us in this room former Speaker of the House Eugene Bennett is in intensive care, suffering from the new virus. These days, when our country is so polarized, it's difficult to imag-ine getting bipartisan consensus on anything, but Gene was a master at that, as well as a compassionate soul. As we try to figure out how to deal with this horrible virus, we need to remember that the case counts are more than just numbers. Sadly, Gene puts a face on it for all of us."

The president gave everyone a moment to take that in. Then, focusing on Goldfield, he asked, "So, where are we?"

"We've seen a rapid increase in cases during the past week," Goldfield related, glancing at her iPad for reference. "There are at least three hun-dred confirmed cases in fourteen states. And state public health directors are anticipating a significant number of additional cases in the next sev-eral days. They are also reporting that states are reacting in different ways, depending largely on the local political environment. Plus, we're seeing a rapid rise in confirmed and suspect cases in Asia and across Europe."

"Not surprising," the president commented, and seemed to throw a quick glance to Stern before looking back at Goldfield. "Where are the greatest num-bers of cases?"

"In large cities, primarily," she responded. "But with the limited amount of follow-up we've been able to do, we think that primarily reflects direct access to the international airports. That's where we believe the virus first entered the country, including Atlanta-Hartsfield, Dallas Fort Worth, New York JFK, and Minneapolis–St. Paul."

"I have a question," the vice president stated. "This new virus came from a part of Africa known to be infiltrated by Al Shabaab and linked to Al Qaeda. What are the chances this is an intentional release?"

"We have no intel to suggest they have that kind of research or lab capability, or connections with the folks who could create a virus like this," Marsh replied. "And there is nothing in the virus's genetic profile that suggests it's been engineered. But we're looking into the possibility of an unintentional leak from one of the labs working with coronaviruses and doing gain-of-function studies."

"That means purposely enhancing a viral agent to anticipate what it might do in nature and preparing medical countermeasures," Sadlock clarified.

"Is that even a good idea, this gain of function, I mean?" the vice president asked.

"It's controversial," Goldfield said. "Particularly now that it's so easy to post genetic sequences on the internet. But that's likely not what's happening here."

"Okay, so what did we learn from the Covid experience that's going to guide us?" the president asked.

"Well," Goldfield said hesitantly, "we never really did a full review as was done after 9/11. Senators Patty Murray and Richard Burr—one Democrat and one Republican—proposed it, but it never went through. Honestly, I think there are a lot of lessons not yet learned from Covid."

"So, what the hell do we do now?" the president asked, his patience clearly wearing thin. "How long before we have a vaccine?"

"To develop, test, manufacture, scale up, and organize distribution—best-case scenario would be four months. But that's no guarantee. A lot of science is trial and error. And remember, that's just to get an Emergency Use Authorization from the FDA. It will take months more before a lot of vaccine is manufactured and actually going into people's arms."

The president surveyed the room, making eye contact with everyone in turn. "We've got a lot of really smart people assembled here. Hell, I appointed most of you myself. So, with what Tamara's telling me, what do we do in the meantime? How do we avoid the mistakes the first Trump and Biden administrations made with Covid?"

Stern shifted in his chair and, in his usual assertive style, said, "We've *got several issues here: There have been calls for closing the borders and inter-national travel terminals. The airlines and the traveling public are very nervous."*

"That would be shutting the barn door after the horse has run out," Goldfield responded.

"You know the Hill is going to want to close the borders, even if it's nothing but political theater at this point." Stern returned a frustrated stare at Goldfield.

"Also," the president continued, "what are we going to do about schools? How about offices, businesses, restaurants?"

"And there's always the hot button of churches, synagogues, and mosques," Brian Caldwell added.

"What are other Western countries doing about things like lockdowns and mandates?"

"So far, pretty much what each one did during Covid," Goldfield said.

"Okay. Which ones were successful?"

Goldfield, Sadlock, the ASPR, and the NIH director looked nervously at one another. Clearly, the others wanted Goldfield to respond. "Well . . . that depends," she began.

"Depends on what?" the president said sharply.

"On when in the pandemic you're talking about."

"I'm not following you."

"Each country, or each of our states, for that matter, will tell you that their approach was the most effective, saved the most lives, and caused the least possible economic disruption in the course of saving those lives. For example, with China's zero-Covid enforced lockdown, they initially had rela-tively few deaths for the size of their population, especially considering that they had a weak healthcare system and an inferior vaccine, and many people, particularly the elderly, were not vaccinated at all. Then after two and a half years, the public would no longer accept the government's draconian mea-sures, and they had to let up, just at the time Omicron — the most infectious variant we'd seen up to that point — was spreading. It was like throwing gas-oline on a fire. We'll probably never know how many people died, but we know it was horrific — at least 1.8 million deaths in the first two months after zero-Covid was relaxed, by one estimate. Still, for those first two and a half years, before Omicron, they did contain the pandemic. But as we all know, the eco-

nomic cost to the country, and to the rest of the world that depends on China's manufacturing, was enormous.

"While Denmark, Norway, and Finland went into their own forms of lockdowns," Goldfield continued, "Sweden went in the opposite direction: not closing down, no stay-at-home orders or mask mandates, and schools stayed open."

"And . . . ?" the president asked.

"In the first year of the pandemic, Sweden's mortality rate was ten times Norway's, and something like eight times Finland's and four times Denmark's. But by the time the pandemic ended, the gap in mortality had narrowed substantially. That's probably because Sweden's vaccination rate reached 78 percent, and the Swedes have a strong social cohesiveness and sense of responsibility for their neighbors. The bottom line? Basically, how well your country or state did with Covid depended a lot less on what measures you took than on when you got hit with surges, what variant was in circulation at the time, what percentage of your population was vaccinated and how many doses of vaccine they'd had, and how many people had natural immunity from previous infection. We have no idea how long this pandemic is going to last. At this point, though, it looks a lot more infectious and virulent than Covid."

"Well, we're not closing down the country," National Security Advisor Marsh declared. "That will never fly!"

"What happens if this is exactly like Covid?" Borges posed. "Are we just going to let it spread?"

"What choice do we have?" Marsh shot back.

"This could be much worse than Covid. It appears to be a MERS-like virus, but much more infectious. And there's likely very little to no protection from having been through Covid," Goldfield said solemnly.

"What about mandates?" the president asked. He turned to Stern. "Gil, you're a lawyer. What powers do we have?"

"Less than you might think," Stern replied. "Decisions regarding mandates are in the hands of state and local officials. We have some authority under Title 42 to prevent diseases from entering the country or moving across state borders, but we don't have much latitude."

"Would that change once we declare a Stafford Act emergency? Or after we issue our Proclamation on . . . What do we call that?"

"Declaring a National Emergency Concerning the SARS-CoV-3 Outbreak."

"Wouldn't that give us the authority to put a mandate in place?" the president asked.

"The Stafford Act allows us to allocate funds toward our response, but it doesn't supersede the states' rights to determine if or how they impose a mandate."

"So what can we do?" asked the president, visibly frustrated.

"All we can do is make recommendations to the states to guide their decisions," offered Stern. "Each governor handled Covid lockdowns differently in 2020, but it appeared that the guidance from the federal level had at least some influence in many states. That doesn't necessarily mean the federal recommendations were correct then or that we should proceed with suggesting mandates now."

"It's like that line from Jurassic Park," *Borges said. "'Your scientists were so preoccupied with whether or not they could, they didn't stop to think if they should.'"*

"At the beginning of Covid, some very good people in public health advocated for mandates to hold down spread so we wouldn't overwhelm the healthcare system while we waited for a vaccine—think of those early days in New York City, for example," Sadlock said. "Others said we should try to reach herd immunity through natural infection, while protecting those who were most vulnerable. As it turned out, herd immunity was an ever-receding horizon. And contrary to what we'd all hoped, the vaccines, while highly effective, lost much of their immunity protection after four to six months. They did a good job of protecting against serious illness, hospitalization—particularly in the ICU—and death, but they didn't prevent everyone who'd been vaccinated from either becoming infected or transmitting the virus to others. So, neither goal proved realistic."

"That doesn't give us much to go on," said the president. "What about schools? As you know, Eleanor and I have a sixth grader and a ninth grader. If I recommend that governors keep schools open, there's no way that we can keep our kids home. The optics would be terrible."

"Well," said Goldfield, "looking back at Covid, we closed schools at the wrong time. Early on, when we shut them down, there was actually a very limited threat of kids getting seriously ill or transmitting the virus to others.

When we got to Omicron, late in 2021, that variant was highly transmissible and much more dangerous to children, but by then the schools were open. With this one, sir, we don't know yet how it will affect kids. So, should we close schools to protect students and teachers, to slow transmission, or both? I wish I could give you a definitive answer, but we just don't have data."

"So, you're telling me you don't know yet whether we should even consider advising governors to close schools? And if we leave schools open, you don't know if I could catch this virus from one of my own kids?"

Goldfield and Sadlock both remained silent.

"You're not giving me much to go on," the president repeated.

"I think we need a plan for protecting the president, the vice president, the cabinet, and senior officials," Sulbarry said. "This is a serious enough threat that we have to be cognizant of continuity of leadership."

"Work it out with the Secret Service, the cabinet secretary, and the congressional leadership in line of succession," the president directed. He stood up, indicating the meeting was ending. "Okay, I need a wish list of everything we need funding for—vaccine development, testing, the Strategic National Stockpile, small-business relief, whatever. Convincing Congress isn't going to be easy this time around. I need an estimate of what we need to take to Hill leadership by tomorrow morning."

"What are we going to say about mandates for schools and businesses? And public gatherings?" Weinberger asked. "We've got to release something."

"Until we know more," the president declared, "the governors are on their own."

BY LATE 1776, THE Continental Army of the United States was facing a two-front war. The visible front was the formidable army sent over by King George III and Parliament to put down the colonial insurrection for independence. The invisible but far more deadly front was the war of disease. In fact, researchers at the Library of Congress have estimated that nine out of ten deaths in those early months of fighting came not from British bullets but from microbes.[1] And of the diseases felling the American troops, the deadliest and most debilitating was smallpox. That smallpox was an old-world blight that commonly killed as many as a third of all those infected only compounded the problem.[2] Many, if not most, of the British troops had already had smallpox and recovered (if you are infected with smallpox and recover, you have lifetime immunity), whereas the vast majority of the Continental soldiers were, as the technical term puts it, immunologically naive.

That is why General George Washington made the agonizing decision to inoculate his army against the disease that has been called the Scourge of History. Beginning January 6, 1777, all the soldiers coming through Philadelphia and Morristown, New Jersey, received the smallpox inoculation.

Washington explained in a February 6 letter to Dr. William Shippen, chief physician of the Continental Army:

> Finding the smallpox to be spreading much and fearing that no precaution can prevent it from running thro' the whole of our Army, I have determined that the Troops shall be inoculated. This Expedient may be attended with some inconveniences and some disadvantages, but yet I trust, in its consequences will have the most happy effects. Necessity not only authorizes but seems to require the measure, for should the disorder infect the Army, in the natural way, and rage with its usual Virulence, we should have more to dread from it, than from the sword of the enemy.[3]

Washington had contracted smallpox as a nineteen-year-old in 1751 during a trip to Barbados with his brother Lawrence.[4] Happily, after several weeks of bedridden suffering, he recovered. But the memory of the illness lingered, as did his appreciation for its severity.

The reason Washington's order was so difficult for him, and controversial to many, was that inoculation at that time was primitive and fraught with danger. It involved making a small incision in the recipient's skin and rubbing it with a thread or piece of cloth contaminated with the virus. The thought was that this means of exposure—contact, rather than inhalation—would invoke an immune response without the high risk of serious illness associated with the normal transmission mode. But if this didn't turn out to be true—if many soldiers became seriously sick or died from the procedure—it would not only decimate his army; it would also present a major obstacle to further military recruitment. And Washington must have figured there was a good chance that if his plan backfired, the Continental Congress would relieve him of his command.

By the end of February, the entire army was inoculated. Fortunately for Washington, the Continental Army, and the nascent nation, the plan worked. Though there were some casualties, not a single regiment was incapacitated. According to a National Park Service history prepared for Boston National Historical Park, "Though it was a controversial action, many historians credit the medical mandate with the colonists' victory in the Revolutionary War and the creation of the United States of America."[5]

The mass inoculation also helped establish the legal principle that certain medical or health-related decisions could be mandated if an overriding public protection or safety issue was concerned.

We will take a deep dive into vaccines—the sharpest arrow in our public health quiver—in Chapter Four. But we take up mandates first because when an epidemic or pandemic hits, it is usually the nonpharmaceutical interventions (NPIs), including the powers of law, that are what we have to work with.

In civilian life, mandates have typically been less about compelling an individual or population to undergo vaccination or adhere to certain directives (such as mask wearing) and more concerned with giving access or services only to those who comply with the mandate. One example is the requirement by school districts for students to be vaccinated against common childhood diseases before beginning class. As early as the 1905 decision in *Jacobson v. Massachusetts* in which the US Supreme Court upheld the authority of the states to enforce compulsory vaccination when deemed to be in the public interest,[6] it has been clear that in order to protect public health and safety, the scope of the state's police power includes the authority to enact reasonable regulations, including

mandates. Similar laws and court rulings proliferated in the following years.

The idea that governmental authorities could mandate adher-
ence to policies to protect public health was already an accepted
legal concept before General Washington's smallpox order. By the
seventeenth century, several European cities had adopted rules
allowing for isolation of the contagious sick in their own homes or
other facilities, and the quarantining of ships and their crews in har-
bors if there was reason to believe they might be carrying communicable disease. These practices soon made their way to the New World
colonies, and by the mid-1700s, major American cities had established public hospitals.[7] John Griscom's *The Sanitary Condition of the
Labouring Population of New York* in 1848 led to the establishment of
the New York City Health Department a few years later.[8]

By the turn of the century, forty states and municipalities had
established their own health departments.[9] This shortly after medical science, led by pioneers such as French chemist Louis Pasteur,
began to accept germ theory and understand the etiology of bacterial diseases like tuberculosis, diphtheria, and typhoid. Pasteur led
the way with a technique for producing rabies vaccine for humans
and immunizing chickens against avian cholera and cattle against
anthrax. Bacteria were identified long before viruses because they
were much larger and could be seen under the microscopes of the
time.

As cities became more populous and the urban poor were concentrated into crowded and unsanitary slums, it became increasingly
clear that the wealthy could not avoid disease by staying away from
the impoverished, and public health was everyone's problem. Massachusetts bookseller and statistician Lemuel Shattuck wrote in
his 1850 *Report of the Massachusetts Sanitary Commission*, "Even those
persons who attempted to maintain clean and decent homes were
foiled in their efforts to resist diseases if the behavior of others
invited the visitation of epidemics."[10] As public health became a
more science-based enterprise, governments relied on established
experts to recommend and make policy. In 1907, for example, the

Massachusetts legislature passed a law that required the reporting of individual cases of sixteen specified diseases.[11]

The 1918 influenza pandemic was a key test of the effectiveness of mandates, including isolation efforts, since the virus was largely spread through crowded conditions in military camps, troop transports, and the trenches of World War I's Western Front in France and Belgium. In addition, the military command had absolute authority to compel compliance with health directives. This was a crucial consideration, because between September and November 1918, influenza and related pneumonia sickened anywhere from 20 to 40 percent of the American Expeditionary Forces. Before the war was over, more American soldiers, sailors, and mariners would die of disease than of battle wounds.[12]

Historian John Barry became a leading authority on public health and the 1918 flu pandemic after writing his seminal and best-selling book on the subject, *The Great Influenza*. He told us that of 120 army bases in the United States, 99 undertook NPIs such as masking and distancing, while 21 did nothing. "Statistically," he said, "there was no difference in morbidity or mortality between those that did something versus those that did nothing." He then qualified the statement by saying that, in general, "The NPIs had some real impact, more so than was thought possible, though sustaining them was difficult.... Sustaining compliance is always going to be an issue—for weeks or months. It's damn near impossible without measures like the Chinese took [during Covid]."[13]

A contemporary article in the November 8, 1918, issue of *Science* confirmed Barry's observation:

> In theory and practise influenza is preventable but it is very difficult to control under municipal and military conditions. It rarely happens that the necessary measures—chiefly isolation—are taken in time. In the present pandemic the disease has, on more than one occasion, been confined to certain wards of hospitals to the exclusion of others. It is not possible as yet to state to what extent it has been restricted in [military] camps. No large camp has escaped it.[14]

As a 1928 retrospective analysis from the Army Medical Department explained, "To be of avail in excluding influenza, quarantine must more nearly approach perfection than proved practicable in the large camps of the war period." The report went on to concede, "When all is said, the best result to be expected from any or all of these [quarantine and mitigation] measures is a slowing of the progress of an epidemic rather than any considerable diminution in the number of cases."[15]

The only places that escaped the ravages of the pandemic were communities that managed, thanks in part to features of their respective geographies, to cut themselves off completely, such as Gunnison, Colorado, and Fairbanks, Alaska. But the restrictions had to be rigid. American Samoa did not record a single case, while Western Samoa, only a few miles away, lost 22 percent of its population to the virus.[16]

An extreme example of the exercise of quarantine authority occurred several years earlier. That is the case of Mary Mallon, an Irish immigrant who came to New York toward the end of the nineteenth century and worked as a cook for affluent families in and around New York City. Over time, it became clear to George Soper, a sanitary engineer for the New York City Department of Health, that Mallon was transmitting typhoid, even though she remained healthy and had no symptoms herself.[17] Soper and his team traced Mallon's employment career and were able to confirm twenty-two cases[18] (by the time of her death, more than fifty cases had been connected to her, including three deaths[19]). Against her will, Mallon was quarantined at Riverside Hospital on North Brother Island in the East River. Except for a five-year period during which she was released but continued to spread typhoid fever,[20] Mallon spent the rest of her life on North Brother, passing away in 1938 at the age of sixty-nine.[21] Technically, this was what we would now call isolation, though at the time it was considered a quarantine. It was after George Soper published an article on his investigation in the June 15, 1907, issue of the *Journal of the American Medical Association*[22] that the sensation-hungry media picked up the story and bestowed Mallon with the moniker she continues to hold to this day: Typhoid Mary.

The Mallon case has been debated in legal, ethical, medical, and public health circles for more than a century. Did the interests of the many not to be contaminated supersede the rights of the individual to her own freedom? Under the circumstances, was she treated well or poorly by the legal and medical establishments? Were there other ways of controlling her actions and preventing infection? Was it right to effectively use her, against her wishes, as a lifelong medical experiment? And what about the more than four hundred others later found in the New York region to be asymptomatic spreaders who were never confined?[23] Where is the balance of rights and responsibilities properly placed? That last question, in particular, is one we are still asking today.

The historical precedent for the use of government-mandated restrictions in influenza pandemics since 1918, as well as the modern legal basis for their use with Covid in 2020, will be relevant to how we deal with the Big One. For the past hundred years, influenza pandemics were the only examples of infectious diseases that can spread quickly around the world and for which we have limited, if any, pre-existing population immunity. Influenza, like Covid and the coronavirus of our thought experiment, is a virus with wings, transmitted through the air simply by breathing. As we've discussed, NPI mandates in military camps in 1918 did not result in overall reduced influenza morbidity or mortality. It's notable that during the three influenza pandemics following the 1918–21 event—1957, 1968, and 2009—no government or public health organization established, or even called for, widespread mandated NPIs.[24] During the summer of the 1957 H2N2 pandemic, for example, state health officials decided to take no public health action and wait for a vaccine to arrive in the fall. At an August 1957 meeting of the Association of State and Territorial Health Officials, it was concluded that "there is no practical advantage in the closing of schools or the curtailment of public gatherings as it relates to the spread of this disease,"[25] because, the participants concluded, there was no practical means of limiting the spread of infection.

In March 2020, as the Covid pandemic was coming into focus, in the United States it was clear that state and federal authorities had the necessary police powers to enact and enforce mandates to protect the public's health. In a 1985 article in *The Hastings Center Report,* Dan E. Beauchamp, then a professor of health policy and administration at the University of North Carolina's Schools of Public Health and Medicine, outlined the balancing of population protection and personal freedom that still pertains today: "The central principles underlying the police or regulatory power were the treatment of health and safety as a shared purpose and need of the community and (aside from basic constitutional rights such as due process) the subordination of the market, property, and individual liberty to protect compelling community interests."[26]

Michelle M. Mello, professor of law and public health policy at Stanford Law School, is a researcher and well-published author on subjects related to ethics, law, and health policy. She sums up the essential issues with health-related mandates: "Two questions to think about when we seek to understand whether mandates are justified because they are effective: (1) Would a mandate cause enough change to justify it? And (2) given gaps between what we are doing and what we should do, is this mandate justified?"[27]

JUSTIFYING THE MANDATE

For twenty years, I led the Minnesota Department of Health's infectious disease epidemiology activity. Our group became one of the most expert such units in the world. For seventeen of those years, I was the state epidemiologist, during which time I testified on many occasions before our state legislature and in public meetings and events promoting the need to expand childhood immunization laws that included newly available vaccines and to limit loopholes that allowed parents to circumvent the law. These vaccines mandated for public school attendance are highly effective in stopping infection, disease transmission, and potential life-threatening illnesses. I also led outbreak investigations that resulted in the state immediately withdrawing potentially contaminated food products from the market and closing businesses that were responsible for disease transmission.

These included outbreaks of Legionnaires' disease associated with contaminated building HVAC cooling towers. Twice I had to order under the authority of the commissioner of health that monkeys kept as pets that had bitten people should be euthanized and tested for rabies (there is no test for rabies that can be performed on a live animal) and herpes B virus, a Simian virus deadly to humans. Although I could not order that if the test results were positive, the humans bitten should receive post-exposure treatment, I strongly recommended it. One such case happened to occur on Good Friday. Animal rights advocates launched a public campaign against the euthanasia order and labeled me Pontius Pilate.

This is all to say that I fully appreciate the critical importance of the state's police power when such action is the only way to protect the public from serious infectious disease risks. No one can claim I'm against the use of public health mandates. At the same time, I have always been acutely aware of the responsibility of those of us in public health whose authority stems from state or local government to balance whatever action we take against any costs incurred, including economic loss, personal liberty restrained, and ideological beliefs overridden. I often emphasized to our team at the Minnesota Department of Health that we had an obligation to never be wrong in our actions. Should we incorrectly identify a party or individual responsible for the transmission of a serious infectious disease, it would take a lifetime to recover the trust of the community. For this reason, I have always advocated that if a government is considering a public mandate, those in positions of authority must have and must share compelling data that the mandate will have, or is highly likely to have, a positive effect, and that the effect has been weighed against any negative effects or consequences, whether medical, economic, or social. The bottom line: A mandate should be employed only if, and only as long as, there are measurable data that it is making a significant difference in morbidity or mortality. A good example of this was the dramatic drop in measles cases, as well as deaths, with the advent of mandated measles vaccination for school attendance.

So, how do we decide when it is worthwhile and necessary to restrict personal freedom or require compliance in the interests of public health? What metrics and values do we bring to that decision? How do we consider the merit of any action recommended by public health agencies and enforced by governmental authority? And how do we monitor the ongoing effectiveness of such decisions to make sure they are worth the costs? In a pandemic like Covid-19, or the Big One, that lasts years, how do we ensure that whatever mandates are put in place can be efficiently updated as we gather more data, and as the virus and our understanding of the science evolves?

UNLEARNED LESSONS

I am not aware of a single pandemic preparedness plan considered by any state, county, or city before 2020 that addressed community-wide lockdowns, mandatory masking in nonclinical settings, and/or closing national borders, as we saw in the early months of the Covid-19 pandemic. All those measures were hurriedly put into place, reactively, without any deep thinking or discussions among public health professionals and policymakers as to what the metrics would be to require such actions, or how and when to end them. Nor was there any attempt to arrive at a consensus about how these measures would meet the ultimate goal of limiting the morbidity and mortality of the emerging pandemic.

Pandemic influenza plans prepared by WHO, the CDC, and other federal, state, and local agencies—including preparations for mitigating a pandemic on the order of 1918—have been insufficient and not effectively integrated to enable a whole-of-government approach to managing the response. Based largely on influenza models and the assumptions that come with them, the plans haven't addressed real-life contingencies and realities of human psychology, economic imperatives, or microbial behavior. And now, having experienced a coronavirus pandemic, we have not improved on outdated influenza pandemic models and still have not seen any detailed analyses that weigh the intended result against the negative impacts or downstream effects of various Covid-triggered mandates or lockdowns.

The 2009 H1N1 influenza pandemic should have been a trial run for testing these measures, or at least a wake-up call. What was different about the early days of the Covid pandemic compared to the 2009 influenza outbreak, which spread to 135 countries and territories in just the first month?[28] Really, nothing. With both Covid and 2009 influenza, ground-zero experiences in Wuhan and Mexico in the earliest days of each pandemic supported the conclusion that the percentage of infected patients who were dying was substantial. These stats made for scary possibilities. Both emerging pandemics were caused by highly infectious, respiratory-transmitted viral pathogens. Ultimately, Covid was lethal to many more people than the 2009 H1N1 virus, but in the early weeks of each pandemic, they looked similar in their severity.[29]

In 2009, we had few mitigation plans for what we'd do as we waited six months for an effective vaccine to be available, and then only in limited supply. And none of the community mitigation plans we proposed contemplated mandates. The CDC recommendations for NPIs at that time focused on voluntarily staying away from workplaces and schools if infected, and decisions about implementing these recommendations were left to state and local health department officials.

The earliest days of the Covid pandemic resulted in an entirely different approach to government mandates: stay-at-home orders versus voluntary exclusion of infected people at work and in schools; border closings; and, eventually, mask and vaccine mandates.

An article published in the April 21, 2017, issue of the CDC's *MMWR* was entitled "Community Mitigation Guidelines to Prevent Pandemic Influenza — United States, 2017."[30] The title sounds comprehensive and confidence-inspiring, and it did contain guidelines for shutting down schools, physical distancing in work settings, canceling or postponing mass gatherings, and other NPIs. But there was nothing that anticipated an intervention lasting longer than a couple of months or addressed what to do if a respiratory-transmitted pandemic dragged on for years. Of note, the guidance did not include any discussion of lockdowns or vaccine mandates. I submit that the public health playbook we used in our response to

Covid showed that little was learned at either the federal or state level from the 2009 H1N1 pandemic.

That won't do when the Big One comes around.

A pandemic that lasts two or three years becomes a sum-total game in terms of case numbers, and in the most likely scenarios, we won't know at the outset how long the outbreak is going to last. Even if you impose a near-total lockdown, as China did, which would be impossible in less authoritarian nations, it has to end sometime. And if the virus is still around and easily transmissible at that point, what has been accomplished?

Mandated restrictions are hardly ever effective for the life of a years-long pandemic. Aside from the difficulty of scientific evaluation, after a point, most people will simply grow weary of them and refuse to comply.

Let's be clear on what we're talking about: Mandates, as practiced in the United States and other Western and democratic countries, do not require you to comply with a particular behavior, such as wearing respiratory protection or getting vaccinated. Rather, like the example cited earlier of school vaccination requirements, they limit various kinds of access for those who don't comply. A government-imposed mandate could limit admission not only to public schools but also to public properties or commercial gathering places like restaurants. A company or corporate mandate could require compliance to be allowed to enter the company's premises. In the past, courts have ruled on such mandates in different ways, depending largely on the circumstances but also, in some instances, on prevailing public sentiments. During Covid, several mandates closing houses of worship were struck down by state and local courts.[31]

How do we apply the array of public health tools—such as isolation, quarantine, social distancing, shelter-in-place orders, mandatory masking, vaccination, and testing—within government police powers and mandates? In the larger public policy context, there has been legal and philosophical disagreement over whether mandates should be used only to protect public health and safety or should also be applied to protect individuals. A good example of this is automobile seat belts, which have been widely mandated even though they are aimed at the safety of the driver and passengers of a vehicle rather than the public at large. The same can be said for motorcycle helmet laws. Should automobile and motorcycle drivers be required to protect themselves whether they wish to or not? Some say this should be a matter of personal freedom. The counterargument is that noncompliance generally leads to greater injury in an accident, which overburdens emergency medical services and hospitals, often at public expense. This is where the vaccine and mask mandates come back into play: If noncompliers become infected because they did not follow mandates or recommendations, they may become a greater public burden or expense.

That burden and expense are not only a question of financial resources. Consider Los Angeles dealing with a 905 percent increase in Covid cases between November 2020 and January 2021 that led to the county's emergency medical services advising EMTs to conserve oxygen in the field and not transport to the hospital any patient unlikely to survive.[32] This is just one example. If we also consider all those whose disease or injury was made worse by waiting to be seen by medical professionals during the waves of Covid infection—including those who didn't seek help at all—and weigh how much worse conditions will be when the Big One hits, the argument for doing anything we can to prevent additional taxing of our already overburdened healthcare system seems compelling. But can mandates accomplish that? And will the public accept them?

Based on more than thirty years of studying, training for, and writing about pandemic preparedness, I've concluded that in the earliest days of the Covid pandemic, scientifically, we were flying by the seat of our pants. Unfortunately, the mandates enacted—border closings, lockdowns, personal respiratory protection (masks), vaccination, and school closings—drew substantial pushback from the public and policymakers and ultimately cost the public health community dearly in terms of trust and goodwill. Making matters worse, looking in the rearview mirror—or the "retrospectoscope," as we sometimes flippantly say in science—we still don't have a scientific basis to recommend any as a one-size-fits-all approach. Studies show *some* effectiveness for *some* of these approaches under *very specific conditions*.

Not only is the evidence of effectiveness for NPIs weak, but the longer we want to keep these interventions in place, the fewer people will be willing (or able) to comply. While there are a limited number of negative consequences to closing schools or businesses for a few weeks or a month, what happens if we're talking about eighteen or thirty-six months?

I'm not saying I would be against mandates (or at least strong recommendations) if or when they could be shown to be both effective and feasible, but I want to see the evidence or persuasive scientific rationale before we start implementing measures simply because they seem like good ideas or will make it look like the government and public health authorities are doing something. Without a rigorous and detailed reporting and evaluation system, we will never know if we are doing more good than harm.

BORDER CLOSINGS

Plain and simple, as was considered in our thought experiment, border closings do not stop respiratory-transmitted viruses—viruses with wings—from spreading quickly around the world.[33] Nor do they give countries that are not yet experiencing widespread community transmission the time to ramp up the speed and effectiveness of their public health, medical, or public policy responses. Politicians should stop advocating for them, and we should stop employing them.

Throughout my career, I have seen time and time again that the immediate political response to infectious disease outbreaks throughout the world is to close national borders. This is almost always a political decision, not a science-based one. When WHO, on January 30, 2020, declared SARS-CoV-2 to be a public health emergency of international concern, the agency made it clear that countries should not close borders, because such closures are ineffective.[34] Yet, by February 27, thirty-eight nations had instituted border restrictions and other measures that significantly interfered with international travel.[35] Eventually, more than a thousand international border closures were put in place.[36]

Once a newly emerging, airborne-transmitted virus with pandemic potential crosses any border, it's already in the process of rapidly going around the world. Each day its global footprint expands, even when it's not yet fully recognized as an emerging pandemic situation. And it's often not clear where the virus is crossing a border or who is bringing it in. For example, our SARS-CoV-3 crossed national borders within and outside Africa before anyone noticed. But even if that was known and the borders were closed to everyone traveling from Africa, you missed all the people who were exposed on flights or in airports sharing air with infected individuals. In the early days of Covid, the United States focused on the virus entering the country via air passengers from China and other parts of Asia,[37] since that's where the novel virus was first identified. Yet in retrospect, SARS-CoV-2 was already spreading around the world and actually making its beachhead in the eastern states, particularly New York City,[38] via infected passengers arriving from Europe, especially Italy[39]—most of whom were contacts of people who had recently been in China.[40] Border closings proved to be, in the words of our scenario's CDC director Tamara Goldfield, shutting the barn door after the horse had already run out.

In the early days of a pandemic, border closures serve only to punish nations willing to be transparent about the epidemiology of in-country virus transmission, illness, and deaths. Border closings often mean immediate cessation of trade, at great economic cost. We want countries to be transparent about emerging outbreaks within their boundaries so that we can face the challenge together, with the best possible data. Border closings provide incentive to hide such information from the world, the exact opposite of what we need to fight a pandemic.

LOCKDOWNS

I was astounded the third week of March 2020 when I saw public health recommendations to governors and mayors that they should respond to the pandemic through lockdowns—more of a reactive tactic than one born of reasoned, long-term planning.[41] Government had not used statewide lockdown mandates in the previous hundred years for influenza pandemics, nor had we prepared for actions like lockdowns. And it showed.

Pandemic influenza plans prior to Covid-19 anticipated the potential need to "flatten the curve," employing temporary efforts to reduce transmission by limiting select activities so that case numbers didn't surge and overwhelm hospital capacity in a given city or region. Lockdowns, in contrast, are community-wide orders to shut down everyday life and much of commerce with the goal of controlling the pandemic. The "flatten the curve" plans did not assume that the overall number of cases would be reduced, but rather that they would be spread out over a longer stretch so hospitals and healthcare personnel would not be overburdened. Hospitals trying to care for five hundred infected patients over ten days would experience emergency crisis care conditions, but if that same number of patients could be spread out over ninety days, healthcare personnel and facilities could better deal with the challenge. For the Covid pandemic, lockdowns were largely state- or local-declared stay-at-home orders. As noted in our thought experiment, these types of government mandates can be imposed only at a state or local level unless related to interstate or international travel.

Many of these lockdowns were implemented with little regard to the length of the pandemic journey we had ahead of us. In April 2020, I made a prediction to the national news media and my public health colleagues that over the course of the next eighteen months we could experience up to 800,000 Covid-related deaths in the United States.[42] At the time, many considered me an irresponsible fearmonger for making such a scary declaration. Tragically, in September 2021, we were well over 700,000 deaths.[43] No one wants to be right about something like that.

The reminder expressed by the president in our scenario, that the case counts are "more than just numbers," paraphrases a truth that haunts me: Each number represents someone's grandma or grandpa, mom or dad, spouse, child, or other loved one. What I wanted people to take away from my estimate was that I foresaw the pandemic playing out for at least eighteen months or more. I believed it was likely to occur in waves over that time, and I didn't rule out the possibility that it might even last three years or longer. I was concerned that the idea that we would be in this crisis for a long time seemed lost on most people. Government responses to the early days of the pandemic seemed like a hurricane response: It will be bad for a short time, and then we can shift quickly into recovery. President Trump publicly stated in the early months that the pandemic would end soon, saying, "It's like a miracle, it will disappear,"[44] and he maintained that position for months.

Though the pandemic did not resolve as quickly as many had hoped, lockdowns were short-lived. A team of researchers at the Blavatnik School of Government at the University of Oxford that tracked lockdowns in the United States found that of the forty states that issued stay-at-home orders between late March and early April 2020, all but one lifted them or heavily scaled back their restrictions by the end of June.[45]

Not only were these lockdowns relatively short in duration, but most were also loose enough to drive a Mack truck through. In my home state of Minnesota, Governor Tim Walz led a public health–supported response throughout the pandemic as well as any governor in the country, and the Minnesota Department of Health responded to Covid as well as any state health department. Yet while Governor Walz mandated a statewide stay-at-home order in mid-March 2020, it hardly represented a true lockdown — 80 percent of the Minnesota workforce was deemed "essential," and thus was not covered under the order![46]

During the same time that governors used their police powers to order retail stores and restaurants to close down, there were other restrictions placed voluntarily to reduce the risk of transmission. For example, professional sports leagues played their games without spectators. and some houses of worship moved to virtual or drive-in services. In some cases, these voluntary actions outlasted government mandates. This has likely contributed to the public's collective misremembrance of long, strict lockdowns that simply didn't occur. And had longer or stricter lockdowns been in place, it is highly unlikely that the public would have complied.

Importantly, for a stay-at-home mandate to be effective, we must know what is happening at the granular city or county level, in near real time. We need to know not only where and how the virus is spreading, but also the conditions in hospitals and other healthcare facilities in that area. Lockdowns must be strategically targeted, and the acknowledged goal should be about healthcare management more than pandemic control. The short-term success of lockdowns in New York City in spring 2020 exemplifies this point, though a number of hospitals there still had to resort to crisis standards of care. [47]

Trying to make this point clear, on March 21, 2020, Mark and I published an op-ed in the *Washington Post* entitled "Facing Covid-19 Reality: A National Lockdown Is No Cure." Our bottom line: "Significantly reducing the number of serious illnesses and deaths would require a near-total lockdown until an effective vaccine is available, probably at least months from now."[48]

In August 2020, Neel Kashkari, president of the Federal Reserve Bank of Minneapolis, and I published an opinion piece in the *New York Times* describing how to "crush the virus until vaccines arrive."[49] At that time, the first evidence of the effectiveness of the new mRNA vaccines was becoming available, and we recommended that whatever we could do to limit transmission of the virus until the vaccine arrived in a few months could pay big health dividends. But much of the public was already fed up with mandates, and our message largely fell on deaf ears. The public perceived that the lockdowns were more widespread and all-encompassing than they actually were. At the same time, the messaging on masks was so muddled as to be all but meaningless, and there was a general perception that the CDC was either not being straight with us or didn't have a handle on the situation itself.

So another lesson is, if, in the earliest days of a pandemic, we put in place lockdown or stay-at-home measures that aren't clearly grounded in public health necessity, people may well resist other public health messages or directives.

Going forward, if government officials are weighing lockdown mandates, there are four considerations that must go into the decision-making:

1. Stay-at-home mandates should be used only when there is reasonable evidence showing that they can be effective. Don't impose a mandate unless it is based on solid science (as border-crossing restrictions almost never are, for example). Consider carefully whether a specific action will actually reduce the community's risk of infection and transmission.

2. Identify clear objectives and communicate them clearly. Don't impose a mandate based on overpromises, like that the pandemic will soon be over. If the goal is to ease the burden on local hospitals and emergency services, explain that.

3. For a stay-at-home mandate to be effective, it must be something the public is willing to do—that is, it must be feasible and acceptable. Mandates will be effective only for limited periods of time. If you ask for a significant change in behavior over the long term, people will not comply. Do not expect to curtail most aspects of daily life for months or years.

4. Strategically target mandates to specific areas based on the facts on the ground, and be prepared to change as the situation changes. Build in flexibility to accommodate those changing conditions. Explain to the public (and repeat often) what specific action you're taking, why you're taking it, and that the action may change as conditions change. Just as important as why and when you are closing schools and businesses is what the criteria/conditions are for reopening them. People need to hear and understand all of it, and you need to state it to maintain credibility.

MASKING

After the discussion in Chapter Two, there should be little doubt about my position on the critical contribution that *the right* protection device can make in preventing respiratory pathogen transmission. With the use of N95 respirators rather than surgical masks, bandanas, or some other loose face covering, we likely could have significantly reduced the number of people who became seriously ill or died in the Covid pandemic.

Despite my strong commitment to promoting respiratory protection, I cannot support the mandatory use of *nonspecific* masks, because when something is mandated, the public is misled into thinking that it is effective—which isn't true of just any random face covering. I saw countless people over the course of the Covid pandemic wearing inadequate respiratory protection, or wearing masks under their noses or over beards.

Early in a dangerous, highly infectious, airborne-transmitted viral pandemic, the only mandate I could initially support would be wearing properly fitting N95 respirators. I would also recommend their use at home if someone in the household is potentially infected or particularly vulnerable. Employing the Precautionary Principle, if we later learn that the primary means of transmission is not airborne, we can quickly back off such a mandate. While I am realistic enough to accept the likelihood that many people would not comply for personal or political reasons, and that those who did comply would not necessarily wear them properly, such a mandate is the only one I know that, if followed, could save lives and significantly reduce serious illness and death.

But as long as WHO and the CDC provide confusing and often incorrect information on respiratory protection, I can't see how a mandate would be effective. I fervently hope that changes before the next pandemic and that an initiative to develop more comfortable, washable, reusable, and consumer-friendly respirators is realized. In the meantime, please don't count on anything less than an N95 respirator to protect you.

VACCINATION

We will get into the science of vaccines in the next chapter, but suffice it to say that Covid vaccines saved millions of lives over the course of the pandemic and significantly reduced the risk of serious illness, hospitalization, and death. This is an indisputable fact. But those vaccines turned out to be good, not great, public health tools. This is a critical consideration as we get into the vaccine mandate debate.

A study conducted by the Commonwealth Fund estimated that from December 12, 2020, through November 30, 2022, Covid vaccines prevented 3.2 million deaths and more than 18.5 million hospitalizations in the United States, saving $1 trillion in medical care costs.[50] Another study, conducted by Imperial College London, estimated the number of lives saved globally as a result of Covid vaccination between December 8, 2020, and December 8, 2021, at 19.8 million.[51] While I am generally skeptical of statistical modeling studies, I think these provide reasonably reliable ranges of lives saved, based on the assumptions used in their models.

But to understand how we should approach the decision to mandate a vaccine, we need to consider four factors:

1. Is the vaccine safety profile determined? If so, what safety concerns need to be addressed in the population for which we are mandating vaccine use?
2. Does the vaccine provide effective protection? If so, what does that protection look like? Does it generally prevent infection and transmission? Does it provide only limited infection and transmission protection but reduce the incidence of serious illness, hospitalization, and death?
3. What is the duration of protection? Is it many years or decades—as we see with tetanus, diphtheria, measles, and small-pox vaccines—or is it like seasonal influenza vaccine, where protection can significantly diminish within 90 to 120 days following vaccination?

4. Does the disease for which we recommend a vaccine mandate cause serious illness and death to the point at which the public would accept the mandate as important for individual and collective health?

In early 2020, we did not have vaccines in development for a pandemic coronavirus, but work dating back to 1989 showed that mRNA technology might be effective in developing vaccines for such a virus.[52] In 2016, scientists from the NIH and Moderna pharmaceutical company began collaborating on a vaccine using mRNA whose design could be rapidly adapted to newly emerging viruses such as MERS.[53] The following year, scientists published findings from studies of mRNA-based vaccines in humans for influenza and rabies.[54] These studies, together with one evaluating a DNA-based vaccine for Zika virus,[55] showed gene-based vaccines to be safe and potentially effective. The results paved the way for the rapid development of Covid vaccines in 2020.

The trillion-dollar question in 2020 was: Would these vaccines protect against SARS-CoV-2 infection? It was answered in dramatic fashion by the end of that year, when a study found that the Pfizer vaccine was 95 percent effective in preventing Covid-19 infection.[56] Another study found the Moderna vaccine to be 94 percent effective.[57] The Johnson & Johnson viral vector vaccine (this and other types of vaccines are detailed in Chapter Four) had a lower initial efficacy of 67 percent against infection,[58] and 77 percent against severe disease fourteen days after vaccination.[59]

When the results of the two mRNA trials were published in December 2020 in the *New England Journal of Medicine*,[60] there was what I would characterize as public health euphoria worldwide. In the United States, we now had highly effective vaccines that, with the support of Operation Warp Speed, would produce pandemic-changing results in just a matter of weeks.

But not all of us shared the euphoria. Previous work with SARS and MERS vaccines, as well as influenza vaccines, raised questions about the durability of this impressive level of protection.[61] Due to the urgency to develop and administer the vaccines, the two mRNA studies had only limited follow-up on extended protection levels among participants. Both studies included two-month follow-up after the second dose of vaccine for half the participants, and up to fourteen weeks for a smaller subset.[62]

Through the first six months of vaccine delivery and into the early summer of 2021, the key focus was to get as many people vaccinated as possible. But in April and May, we began to hear of an increasing number of Covid cases among individuals who had been vaccinated four or more months before.[63] We refer to these as breakthrough cases.

In the subsequent two years, an increasing number of studies demonstrated vaccine protection was substantially reduced as the variants of SARS-CoV-2 emerged,[64] even with a newer, bivalent (covering two strains) version that was rolled out in 2022.[65] In the earliest days after vaccination, there is protection against serious illness, hospitalization, and death. But the level of protection is a growing concern. For example, one of the most informative studies, conducted by the CDC and five healthcare organizations,[66] found the effectiveness of the bivalent vaccination against hospitalization dropped from 62 percent one week after vaccination to 24 percent at four to six months for those with healthy immune systems. For participants with impaired immune systems, the effectiveness was 28 percent at one week and only 13 percent at four to six months. Investigators found that for those who'd had only the original vaccine two-dose regimen and no booster, effectiveness against hospitalization at one year was 21 percent for those with healthy immune systems and 3 percent for those with immune-compromised conditions.

The public had limited awareness of the vaccines' deteriorating levels of protection, though breakthrough cases were more frequent and obvious over time. Because Covid vaccines provide critical, though time-limited, protection against serious illness, hospitalization, and death, I do strongly support routine Covid vaccine doses every six months, if possible, particularly among those sixty-five and older and those with impaired immune systems. But this is a very different position from supporting a vaccine mandate.

Let's look at the four factors to be considered in the context of Covid-19:

1. The vaccines do meet the standards for safety, and vaccine regulatory agencies around the world have approved them for use.
2. They provide some effective protection, but with limited impact on transmission, unlike mandated childhood immunizations.
3. Durability (long-term protection) is limited.
4. As for whether the disease causes serious illness and death to the point at which the public would accept the vaccine mandate, that's a hard one to call. If you ask people who lived through the pandemic's early, crisis stage in New York City, most would probably answer with an emphatic "Yes!" But people in rural areas that didn't initially see waves anywhere near that level may well respond differently.

Thus, the answers to our four questions lead to one conclusion for our test case: Even though the SARS-CoV-2 virus itself causes enough serious illness, hospitalization, and death to make efforts to combat it worthwhile, the protection from the vaccines available makes it hard to support a case for mandating vaccination with a similar vaccine in future pandemics caused by corona or influenza viruses, until or unless we have some confidence that protection is lasting and the vaccine actually prevents or seriously diminishes transmission.

We don't yet have that magic bullet, despite the initial euphoria.

During the past decade, the challenge of vaccine hesitancy has become an ever-growing concern among the public health and pediatric medicine communities. With the proliferation of social media platforms and the rise of anti-vaccine groups, often supported by celebrities, we've seen a rapidly evolving movement around the world pushing for the elimination of mandated childhood vaccinations for school attendance. Many of us in public health have feared the anti-vaccine movement would use the growing social anger over Covid vaccination programs to fuel that fire.

A Harvard T.H. Chan School of Public Health study published in June 2023 provided a comprehensive analysis of twenty-one polls conducted before and after the height of the Covid-19 pandemic. The study's findings bring to light the complicated relationship between vaccines, parental choice, and public health. It shows that parents are changing how they think about school vaccination mandates, but not because they've lost faith in the vaccines' safety. They are pushing back against the idea that they don't have freedom of choice regarding their children. Gillian SteelFisher, PhD, the lead author of the study, explained, "Public support for these school requirements dropped by 10 to 12 percentage points between 2019 and 2023. . . . This left about one-quarter of US adults (25 percent to 28 percent) opposed to these vaccine requirements by 2023, which is the highest level of opposition to routine childhood vaccination in recent history."[67]

Given the major pushback against vaccine mandates, we really need to ask ourselves if "the juice is worth the squeeze" to mandate vaccination in the next pandemic if we don't have substantial data on the vaccines' long-term protection against infection, illness, and infectiousness.

I can already hear the naysayers proclaiming, "Osterholm has seen the light! He's finally on our side about free choice!" Please don't misinterpret my position. I am absolutely in favor of vaccination, including mandated vaccination, but I always want mandates to be evidence-based. It is nothing less than a tragedy that we are seeing cases of debilitating and even fatal illness from preventable diseases like measles, polio, and others once effectively eradicated in the United States and many places worldwide because safe and lasting vaccines are being shunned.

SCHOOL CLOSINGS

Perhaps the most contentious societal issue that came up during the Covid pandemic was school closings, whether declared by political leaders or school officials. We need to consider it carefully in preparing for the Big One.

As a grandfather of school-aged children, throughout the pandemic I was certainly invested in ensuring my grandkids' safety and education, and their parents' ability to continue working. I was like millions of others struggling with these issues around the world. Along with protecting children, we had to weigh the health of parents and grandparents, as well as faculty, administration, and staff, and their family members—particularly any who had underlying health issues that put them at increased risk for serious illness. We were also concerned about the lost educational opportunities with closed schools and distance learning. Students from low-income environments where access to internet-based education was not available represented another aspect of the way the pandemic exacerbated inequities in our society—and remember the many who regularly received much of their nutrition in school. Parents had serious concerns about their children being at risk for infection in school, but many nonetheless insisted their children needed to attend school both for their education and because of the economic reality that a sizable percentage of parents were simply unable to stay home from work to care for their homebound kids and had no alternative care providers.

My fifty-year career in public health has taught me one critical lesson in risk assessment and communication: When children are involved in an outbreak or other infectious disease crisis, expect worried parents to find any risk to their children unacceptable. That might seem to contradict the previous statement about many parents protesting the closing of schools during Covid, but what it more likely demonstrates is the impossible choices many parents found themselves stuck between: understandably, they often want business as usual, but with their children completely protected, which is impossible. Public health policy almost inevitably involves a trade-off between risks and benefits. These days, all the political and social media voices can blur the facts that feed into analysis of those trade-offs. We must plan now to avoid another pandemic school meltdown.

As we've suggested, almost no one—not school officials, faculty, and staff; parents; public health officials; academic researchers; politicians; or news reporters—knew that the first year of the Covid pandemic would likely not even be the midpoint; it was kind of like calling the winner in a baseball game after three innings of play. As it turned out, the results of studies attempting to define the risk of infection for children in schools in the first year of Covid did not represent what that risk would be for the duration of the pandemic.[68] Because the early rates of infection were relatively low, school officials, educators, and researchers greatly underestimated the evolving risk with the arrival of variants like Alpha, Delta, and Omicron over the next two years. These variants were more infectious and caused more severe clinical illness in kids. Most regions of the country saw increasing surges of pediatric hospitalizations in May 2021 (associated with the Alpha variant), September 2021 (coinciding with the fall school opening and the arrival of Delta), and January 2022 (when we saw the major surge of Omicron).[69]

One way to consider this changing risk over time is to examine deaths due to Covid in children in the United States. In the first three years of the pandemic, there were 1,559 pediatric deaths recorded: 199 in 2020, 612 in 2021, and 748 in 2022.[70] Eighty-seven percent of the deaths in children occurred in the second two years of the pandemic—but by then, schools that had closed in 2020 were back open! As our fictional CDC director Tamara Goldfield observes in our thought experiment, "Looking back at Covid, we closed schools at the wrong time."

The study that I believe provided the clearest and most compelling data on the risk of infection in children related to school attendance, and the role that these children played in household transmission, involved 166,170 US households with adults and children, using smart thermometers, conducted from October 2019 to October 2022.[71] The authors concluded that more than 70 percent of household transmission in homes with adults and children was from an infected child. When schools reopened in fall 2020, children contributed more to household transmission when they were in school and less during summer and winter breaks. This pattern was consistent for the two consecutive years of the study. Researchers in the United Kingdom found similar pediatric case trends.[72]

In short, as we've just outlined, all the debating about the lower risk of pediatric cases in the first year became essentially irrelevant to our experience in the second and third years of the pandemic. If there was a case to be made for closing schools based on close to real-time data, it would have been later in the pandemic, rather than earlier.

When I emphasized in 2021 and 2022 how the changing epidemiology of pediatric Covid cases over the first three years of the pandemic altered the risk to school-aged children — that it was relatively low in the beginning of the pandemic but increased with later variants — I was often accused of being scary and a voice for the teachers' unions. I received almost a thousand emails and letters challenging my statements and referring me to two articles disputing my conclusions about the changing role school-related transmission played in the evolving pandemic. The first, a November 19, 2020, opinion piece that appeared in the *Washington Post* entitled "We've Figured Out It's Safe to Have Schools Open. Keep Them That Way," was written by Danielle Allen, the James Bryant Conant University Professor at Harvard, and Ashish Jha, the dean of the Brown University School of Public Health, who later served as White House Covid-19 response coordinator.[73] The second article was published in *The Atlantic* on October 9, 2020, by Emily Oster, a Brown University economics professor, entitled "Schools Aren't Super-Spreaders: Fears from the Summer Appear to Have Been Overblown,"[74] which caught widespread national media coverage. It's hard to promote public health messaging when it contradicts notable academic and media-recognized individuals like Danielle, Ashish, and Emily, even when they are wrong.

When I heard parents demanding that schools be closed, I would often inquire what their kids were doing when they were not in school. The frequent response was that they were playing with classmates at someone's home. While I grant you the risk of becoming infected with Covid may be lower with only fifteen friends rather than hundreds of classmates, being or not being in school does not in itself mean children are at greater or lesser risk of being exposed and infected. This was brought home to me in 1995 when I led an outbreak investigation of ten cases of bacterial meningitis in a Minnesota high school and the surrounding community. One student had died. Parents were panicked and wanted the school closed. The primary mode of transmission of the bacteria causing the meningitis was through saliva exchange, such as kissing and sharing sodas. We urged the community not to close the school, as we could have more effective supervision of students and prevent those activities that would result in shared saliva. The concerned parents prevailed, however, and the school was closed. The next day I took a picture of twelve girls all sharing the same soda cup and straw at a local restaurant. We must never forget that while schools can be significant amplifiers of transmission, they are not the only places students are exposed to infections.

And there is another point to consider. During the Omicron surge in the United States in late 2022 and early 2023, many schools in the worst-hit areas had only a skeleton crew of teachers, staff, and administrators available because so many were home sick. Without adequate staffing, school attendance poses other safety risks for children.

What this means is that all factors must be evaluated in making a decision, school closings should be kept to the shortest reasonable amount of time, and it is highly unlikely that a one-size-fits-all approach will work on a national, or even regional, level.

To date, there also remains a missing connection between

keeping schools open and how vaccines can play a role in that effort. As of late April 2024, only 14 percent of children six months to seventeen years of age were up-to-date on their Covid-19 vaccinations.[75] Even if the vaccines provide only limited protection against an infected child transmitting the virus in school, they and subsequent boosters could be insurance for almost an entire school year against developing serious illness or dying, as well as the risk of developing long Covid.

The same cannot be said for masks. Several studies claim that masking significantly reduces school-related virus transmission.[76] However, our CIDRAP team carefully reviewed and critiqued these studies, and not one held up to rigorous scrutiny from a methods or outcome perspective.[77] We have already examined the challenges of using N95 respirators in everyday life for adults. There is simply no way younger kids will effectively use a currently approved respirator. Instead, we see them using loose-fitting cloth masks, which offer virtually no protection. Further, students required to wear their masks in the classroom are allowed to take them off during thirty or more minutes in a crowded lunchroom. How can anyone not understand that the virus doesn't stop spreading because people are eating?!

So, what should we do? I believe we need to consider a "snow day" approach to school closures during pandemics, similar to CDC guidance on school closures during an influenza pandemic. But this guidance was not followed during the actual flu and Covid pandemics. Schools should not automatically close as soon as a pandemic emerges. The decision to close should be based on student absentee rates, the number of sick teachers and staff, and local healthcare capacity. As necessary, take off a week or even a month, but constantly evaluate the data and the facts on the ground.

School closures can be thought of much like other lockdowns: They may not prevent kids from getting infected at some point during the pandemic, but they can slow transmission and help to flatten the curve for a limited period, staving off a major surge in cases.

Much of this feels like common sense, but that is often in short supply when people are sick and dying of something unknown and the public is calling for answers we don't yet have. In considering mandates, we can go all the way back to the era of Dr. John Snow, and even before, when public health and disease mitigation efforts derived mainly from observation rather than known science: Mandates, directives, and recommendations have been implemented more smoothly and been more effective when they conform to existing social values about the logic and reasonableness of the stated goal, and when their impact can be measured so that the public buys into the objective.

TAKEAWAYS

1. Any mandate must be based on the best evidence available. The temptation to act so the public sees you are doing *something* must be resisted. Like all hygiene theater, actions that will not affect infection, transmission, and health outcomes will be disruptive and harmful to long-term public cooperation.

2. "The juice must be worth the squeeze." If you are asking people to do something—whether not holding their annual huge gathering of family and friends for the holidays or getting vaccinated—make sure you have compelling evidence that their action will have long-term, specific benefits to them and society.

3. With any mandate, there must be a practical way to enforce it. With masking and vaccination, for example, the message must be clearly conveyed: *No one will force you to do X, but if you don't do it, you can't have access to Y.*

4. Honest, clear communication is critical. We have dedicated an entire upcoming chapter to this topic because it is so vital,

and it is especially important in public discourse regarding mandates.

5. In future pandemics, we should not consider school lockdowns or closings as the most effective response. Rather, these should be implemented for limited periods only, like snow days, to flatten the curve or get past a limited period of high community-based transmission.

CHAPTER FOUR

MEDICAL COUNTERMEASURES

For a successful technology, reality must take precedence over public relations, for Nature cannot be fooled.
— Richard Feynman, PhD, Nobel Laureate in Physics

[Pandemic Month 4]

"Today, a special edition: The new coronavirus pandemic. It's now just about everywhere, with death rates in some places over ten percent. How are we going to survive the most devastating health crisis in our collective memory? As the pandemic unfolds around the world, it appears that it may be a much larger catastrophe than Covid. What are the president and foreign leaders doing to mitigate this disaster? My guests include leaders from the world of public health and public policy. And we have reports from around the nation and the world as hospitals are having to initiate triage practices and health-care systems are overwhelmed to the breaking point. Welcome to Sunday. It's Meet the Press.*"*

"From NBC News in Washington, the longest running show in television history. This is Meet the Press, *with Jonathan Goodwin."*

"Thanks for spending part of your Sunday with us. While some are denying its very existence, others are saying that with SARS-CoV-3, we may be facing a crisis that has the potential to rival the 1918 influenza pandemic—the worst public health disaster in modern history. What we do know so far is this:

"From an initial cluster of hot spots around the world, mainly centered in east central Africa, Turkey, northern France, and here in the United States, the virus appears to be spreading rapidly and in unpredictable patterns. Therefore, contact tracing—a traditional tool of public health practice—will not be effective. The same turned out to be true during the Covid-19 pandemic. With the current levels of infection, it really doesn't matter anymore who anyone got it from. That genie is out of the bottle, that train has left the station, or pick your own metaphor.

"But now that tests are widely available, we can begin defining the scope of the pandemic. As of Friday, the last day for which we have figures, the cumulative total of cases in the US exceeds ten and a half million, with 1,101,120 deaths reported. Global cases number around sixty-eight million thus far, with about eleven and a half million deaths. For comparison, at this point in the Covid-19 pandemic, 1,761,503 Americans had contracted the virus, and 103,700 of them had died. Worldwide, there had been 5,817,385 cases and 362,705 deaths, which means that at this point, not accounting for underreporting, we have at least two and a half times more SARS-3 cases reported than SARS-CoV-2 and about four and a half times more deaths. The 1918 flu epidemic had a mortality rate not quite as high as what we're seeing from the current virus, and before it was finished in 1921, that pandemic killed as many as one hundred million people in a world with less than one-fourth the present population. So it is no exaggeration to say that we are living in a world of hurt.

"Though schools remain open in most places, as an added precaution, the president's two children are being homeschooled at the White House. And we can report that Arnold O'Malley, the secretary of education, is hospitalized in serious condition, and at least ten members of the Senate and House of Representatives have tested positive and are hospitalized. All this as Washington is still mourning the loss of former Speaker of the House of Representatives Eugene Bennett. We are told there will be a public memorial service, but not while the virus that claimed the life of the great statesman is still running rampant.

"We checked out several of the most vocal naysayers, who are claiming that the numbers I've just presented are made up, inflated, or misinterpreted as actually resulting from other diseases and conditions; but frankly, we were not comfortable with their credentials or credibility for this program. But my first guest certainly has the credentials. Before being appointed secretary of the

Department of Health and Human Services, Dr. Anne Sulbarry was a well-known virologist and vaccinologist who had her own lab at the National Institutes of Health; she was also a visiting professor at the Johns Hopkins University School of Medicine and the Hopkins Bloomberg School of Public Health. She comes to us from her office in Washington, DC. Dr. Sulbarry, welcome to Meet the Press."

"Thanks for having me, Jonathan."

"We'll get to your work on vaccine development in a moment, Doctor. But first, could you tell us what people can do to protect themselves right now?"

"At the moment, there is a lot we still don't know about this virus, even though it resembles the SARS-CoV-2 viral variants in many respects. We do, however, strongly believe at this point that the virus is transmitted through the air, often by extremely small, aerosolized particles, so the only thing I can recommend with complete assurance is that everyone wear a well-fitting N95 respirator face mask or equivalent whenever they're in a situation in which virus may be present. This is particularly important indoors, or on public transport of any kind. Other than that, I would say just use common sense: Try to maintain a reasonable distance from other people, don't make unnecessary trips, and test often if you're planning to be around other people, as I know you do in the studio. The goal here is not necessarily to halt the virus in its tracks, because based on our previous experience, I doubt that could happen. But if we can keep enough people from catching the disease, becoming seriously ill, and needing hospitalization, even intensive care, before we have vaccines or proven medical countermeasures like antivirals, then perhaps we can keep the healthcare system here and in other parts of the world from straining beyond the breaking point. We, like most other nations, have no surge capacity in our systems. We're already desperately short of doctors and nurses, as well as critical drugs and ventilators, so I fear that the whole medical establishment could come crashing down if we don't do something to mitigate the immediate effect."

"And what about vaccines, Dr. Sulbarry? Are we close to having a vaccine for this virus?"

"I believe we are making substantial progress with several different vaccine platforms, including the mRNA approach that people know from Covid-19, as well as the traditional adenovirus vector vaccine and others. The point is, we're not depending on a single approach; we're trying to see what would

work best for this pandemic, in terms of efficacy, number of doses needed, how scalable it is, whether it requires a cold chain transportation and distribution system, things like that."

"I know this is difficult to predict until you've actually come up with something and it's gone through clinical trials, and I won't hold you to it, but how long would you estimate it's going to take before we have a reliable and effective vaccine? And then, after that, how long will it take to have enough vaccine to dispense in quantity?"

"If we're lucky, Jonathan — and unfortunately, luck does play a significant part in all this — I'd say we're still looking at several months at least before we see the first shots of an approved vaccine in arms. Now, through organizations like Gavi — "

"And just to clarify, Doctor, what is Gavi?"

"Oh, yes, Gavi — formally 'Gavi, the Vaccine Alliance' — is an international public-private partnership formed to bring increased vaccine and immunization access to lower-income countries. And through organizations like Gavi, we'll be working to ensure an equitable distribution around the world, since with a pandemic like this, the old cliché is true: No one is completely safe until everyone is completely safe. But still, it will take several months, at least, to scale up manufacturing at facilities around the world, so we'll definitely be looking closely at what other countries can produce of their own vaccine candidates, particularly China, India, and the UK, which currently have significant production capacity."

"And once we have this vaccine — if we have it at all — how long will the protection last?"

"That's also hard to say. As you remember from Covid, the initial mRNA vaccines were highly effective, and we were all hopeful that it would be one or two shots and then done. But as we saw, immunity fell off fairly significantly after a period of months. For those at highest risk of serious illness, a booster dose every six months likely will be necessary. Some vaccines, like the measles shot, essentially confer lifelong protection, and others, such as the polio and smallpox vaccines, provide long-term protection. But each virus is different, and it will take several months for us to know how long-lasting the protection will be with our new vaccines."

"What about a 'universal vaccine' against SARS-CoV-3 that would be

effective against all forms of the virus and so would be less likely to lose its effectiveness?"

"You're actually talking about two different things, Jonathan. With both influenza and coronavirus, researchers have been working for years — decades in the case of flu — to come up with a vaccine that targets the conserved parts of the virus — the components or proteins that don't change from one variant or replication mutation to the next. A so-called universal flu or pan-coronavirus vaccine would be a true game changer. But so far, we haven't come up with an effective one for either virus. That doesn't mean we won't, or we shouldn't keep trying, because the result would be worth many, many times whatever it cost. It would save untold millions of lives and trillions of dollars around the entire planet.

"The second part of your question is in some ways an even more difficult one. If we came up with a vaccine that was highly effective and didn't have to be altered each time the virus changed, presumably it would confer more lasting immunity. But we really couldn't be sure of that until we saw it in action. I'm sorry I can't give you a more definitive answer."

"Understood. But until effective vaccines arrive, assuming they do, do we have any effective drugs — antivirals — to treat people who are seriously ill?"

"Well, we have the drugs we had for Covid, such as Paxlovid, remdesivir, and molnupiravir. But our clinical experience to date suggests they don't work as well against this new coronavirus. There were some promising studies undertaken during the NextGen initiative to develop a new generation of Covid vaccines and antivirals, but with the short window for funding and no master plan for research and development, they didn't get very far. I wish I could say we had more in our arsenal right now, which only underscores the effort we need to make to develop more soon."

"Madame Secretary, with all of these hospitalizations and deaths, it still does seem that a fairly sizable proportion of those exposed do not get sick, or at least only mildly so. What do you say to those people who feel that since they are not old or immune-compromised, they don't have to worry about hospitalization or death and that, to them, SARS-CoV-3 would be like nothing worse than a bad cold?"

"I would say they should think of others they might infect who would get seriously ill, those either living in the same household or in their immediate

*environments. Keep in mind, too, that it's not just the elderly or the immune-
compromised we have to worry about. We are seeing a greater number of children getting seriously ill and even dying compared with what we saw with
Covid. So, I would say that they can hope that the virus is nothing worse than
a bad cold, for them or anyone they might infect, but hope is not a strategy."*

THE PUBLIC HEALTH AND medical tool kits potentially available at the
beginning of a pandemic fall into one of two categories: nonpharmaceutical interventions (NPIs) and medical countermeasures (MCMs).
In Chapters Two and Three, we described the challenges of employing NPIs, both from the perspective of a pandemic that may last years
and in regard to how effective they are in reducing morbidity and
mortality. But they are the first line of defense when a pandemic virus
with wings emerges. The second line of defense is MCMs. These
include biologics (vaccines, blood products like convalescent plasma,
and antibodies), drugs (antimicrobials, antivirals, and those that
reduce damage due to inflammation), and devices (diagnostic tests,
personal protective equipment, and ventilators, which are most often
regulated by a country's government agency that has been assigned
such oversight. In the United States, this agency is the FDA. In Europe
it's the EMA, the European Medicines Agency.

Vaccines are MCMs that have the most potential to reduce the
health impact of a pandemic. In an ideal world, having a broadly protective vaccine with long-term immunity against most of the influenza subtypes and novel coronaviruses likely to emerge, and the
ability to administer such a vaccine globally as part of routine immunizations, could mean greatly reducing the health and economic
impacts of the next pandemic. Vaccines widely available early in an
emerging pandemic could also have a dramatic impact on the length
of the pandemic.

As we stare down the prospect of the Big One, this is one area in
which we are falling short—frighteningly so. I'm especially concerned that there remains substantial confusion among the public
and even international and national public health, research, and
philanthropic organizations as to what constitutes a pathogen of

pandemic potential. This has major implications for the prioritization of funding support for MCM research, development, manufacturing, and stockpiling for these pathogens.

For example, in May 2020, Gavi, the international vaccine alliance, published a document entitled "10 Infectious Diseases That Could Be the Next Pandemic."[1] The list included diseases that can certainly cause substantial morbidity and mortality on a regional basis but that have no potential to become worldwide epidemics with respiratory spread. Other governmental and philanthropic organizations have developed similar lists.

In 2023, the NIH's National Institute of Allergy and Infectious Diseases published a strategy "to accelerate the discovery, development, and evaluation of medical countermeasures against new or previously unknown pathogens of pandemic potential."[2] Ironically, influenza viruses and coronaviruses were identified as having a high potential for future pandemics, but they were excluded from additional consideration because the NIH had already allocated resources to develop MCMs for these diseases.[3] Viruses from ten other families were considered pathogens of pandemic potential. While these diseases—among them dengue, Lassa fever, chikungunya, and hantavirus—can cause serious regional disease outbreaks, none of the ten viruses the NIH identified as pathogens of pandemic potential are ever likely to actually cause a pandemic because based on current studies, none of them spreads through airborne transmission. Don't get me wrong, these diseases can be devastating on a local or regional basis and deserve major research support. But we must not take our eye off the priority efforts needed to address MCMs for influenza viruses and coronaviruses.

Each of the last four influenza pandemics—1918, 1957, 1968, and 2009—has offered valuable lessons about maximizing our response. But I fear that just as the previous influenza pandemics' teachable moments were quickly forgotten, the same will be true for Covid. By midsummer 2023, the world was quickly moving on, with limited interest in a "hot wash" of what happened and what we should have learned. You will see such a postmortem evaluation in the scenario in Chapter Eight.

The Covid vaccines, which arrived in the first year of the pandemic, were critical in saving lives and reducing severe illness, hospitalizations, and deaths, but they turned out to be good but not great. And they still took a year to deliver. What would the Covid pandemic have looked like if we had been able to deliver great vaccines in that first year? Ones that, after two doses, protected a person from SARS-CoV-2 infection for years, regardless of what variants emerged? Similarly, imagine if we'd had drugs that, if taken by most ill people in the first days of illness, resulted in rapid recovery, even for the immune-compromised. And finally, what if we had had tests that were reliable and readily available shortly after the virus was first detected? While billions of federal dollars were provided to lab-testing companies to develop Covid tests and make them widely available, far too much of this money was wasted, with companies not delivering tests or delivering tests so unreliable that it was almost like flipping a coin.

We have to do better in preparing for the Big One.

VACCINES

The origin story of vaccines is among the most celebrated in all medical history. Edward Jenner, born in 1749, was a London-trained physician practicing in the rural Gloucestershire region of southwest England. For years he had heard anecdotal stories and firsthand personal accounts of how milkmaids seemed to be protected from the deadly smallpox scourges that periodically ravaged the population. He even heard one dairymaid declare, "I shall never have smallpox, for I have had cowpox."[4] Cowpox, which the women contracted from milking infected cows, produced blister-like pustules on the hands and other body parts. These pustules could be painful and unsightly, but the disease was far milder than smallpox. Even though he didn't know what caused cowpox or smallpox, which only infected humans, Jenner theorized that the milder disease was close enough to smallpox that exposure might be what was protecting the milkmaids—or, in our modern parlance, conferring immunity.

This wasn't a completely novel idea even at the time. Jenner was building on centuries of a practice known as variolation, as General Washington required for his troops, in which infected material from a smallpox pustule was scratched into the skin of a healthy person, with the idea that a small exposure would build up resistance to the deadly disease. This approach, introduced long before viruses and other microbial agents were even understood or identified, was an example of the first principle of what later became the science of epidemiology: *observation*. Even if you don't understand why something is happening, recognizing a pattern, and working out a plan based on that recognition, is one of the guiding precepts of public health.

The problem with variolation was that it was imprecise and could be dangerous in itself. The virus in the material that was used produced symptoms and was making people sick, and the procedure had a fatality rate of 2 to 3 percent. While this was much lower than the fatality rate of smallpox when acquired naturally, it still wasn't a very attractive risk.

To improve on variolation, Jenner put his observation and theory into practice in May 1796, taking pus from a cowpox blister on the hand of milkmaid Sarah Nelmes and scratching it into the arms of James Phipps, the eight-year-old son of his gardener. After experiencing a fever and not feeling well for a short time, James recovered. Jenner then injected him with pus from an actual smallpox lesion to see if he was protected, and the boy remained disease-free. Today, of course, such an experiment would be ruled out as going against all medical ethics. It is unknown whether young James consented to the procedure, but given his age, it wouldn't have qualified as informed consent. Fortunately for him, Jenner, and the early history of vaccination, the experiment was a success. Had it not been, it is likely we would have heard nothing about it. We do know that Jenner and Phipps remained friends and Phipps attended Jenner's funeral in 1823.[5]

As with so many other events and breakthroughs in medical history, skepticism outweighed general acceptance. When the Royal Society in London rejected a paper Jenner wrote describing his ongoing experiments, he self-published a booklet entitled *An Inquiry into the*

Causes and Effects of the Variolae Vaccinae, a Disease Discovered in Some of the Western Counties of England, Particularly Gloucestershire and Known by the Name of Cow Pox.[6] Despite the unwieldy title, it may be the most significant self-published manuscript of the modern era and is one of the classic texts in the annals of medicine. Mark and I got to see an original copy of that seminal work in the den of my mentor, the late Dr. D.A. Henderson, in his home near the Johns Hopkins University campus in Baltimore. It was presented to D.A. in gratitude for heading the WHO's international effort over two decades to eradicate smallpox from the face of the earth, one of the greatest public health achievements and gifts to humanity in world history.

Jenner decided to call his new technique "vaccine," based on *vacca,* the Latin word for cow.[7] We could easily write an entire book on the history of vaccination up until now, but suffice it to say that Jenner's basic concept remains in effect to this day: Familiarize the body with a certain pathogen in such a way that it can prepare a defense without having to suffer the actual disease.

Of all the infectious diseases that vaccines aim to target, respiratory-transmitted viruses remain our greatest infectious adversary when it comes to pandemic potential. Practically speaking, there are two kinds of viruses with wings, but only one has the potential to cause a rapidly emerging pandemic. Some highly infectious respiratory viruses like measles experience very few genetic changes over time, so once you've been infected or vaccinated, you become a member of the "protected class." On the other hand, influenza and coronaviruses are highly infectious and experience frequent and rapid genetic changes. Therefore, previous infection or vaccination doesn't mean your immune system will be able to protect against new, mutated versions.

To reduce the likelihood or severity of a future pandemic, we need to prepare our immune system, both the body's detective agency and its hit squad, so that it can provide durable and broadly protective immunity. That's where game-changing new vaccine technologies come in.

Here is a vastly simplified account of the immune response. The "street agents," called dendritic cells, look for foreign interlopers. The dendritic cells inform the helper T cells, which seek out B cells, and in the cases of viruses, together they hunt for the invaders and find elements of the virion surface that they can latch on to. The B cells, now activated, churn out proteins called antibodies that "surround" the virion and then attach to the infectious agent, "calling in" even more immune cells in the attempt to destroy the invading force. And T cells play a role in destroying the virion. When we have a fever and/or feel ill, this search-and-destroy mission is taking place.

To take the detective analogy one step further, once the "culprit" is identified, pursued, and killed, a "case file" of the crime remains in the form of memory B and T cells. If the same or a similar antigen—the term we use for any foreign substance that triggers an immune response—is once again detected in the body, the memory B cells will ramp up to produce more antibodies for the memory T cells hit squad.

Now, the problem is that, as we all know, the good guys don't always win. Some viruses, such as rabies and Ebola, to name just two, are so powerful that they can kill off a high percentage of their hosts. Ebola kills 50 to 90 percent of the people it infects, and rabies has a mortality rate of nearly 100 percent in those who don't receive post-exposure prophylaxis treatment after the bite— usually a series of vaccine shots and rabies immunoglobulin. And for numerous viruses, including those that cause common colds and seasonal flu, even though the good guys usually win, the struggle can make the infected person pretty miserable and take them out of commission for the duration. So we need other weapons to help the immune system win the fight.

It is also important to remember that just as different viruses have varying degrees of ability to infect us, human immunity is not an equally distributed quality, and it is something that can change within a given individual over time, based on any number of life events and variables. Some people are at greater risk of getting infected, becoming seriously ill, requiring hospitalization, and dying from any disease—for example, the elderly, those with underlying health conditions that make them more susceptible to severe illness, and people whose immune system is compromised by a chronic disease or treatment for another condition. Providing them with protection from a virus that might not be so dangerous for a young, healthy person with a robust immune system may mean the difference between life and death.

That's where vaccines come in. A vaccine may be thought of as a way to hand over a criminal's case file to the security guards of the body before the criminal enters the premises so the immune system is given a target to identify ahead of time. Then, if the actual intruder shows up, even in disguise (i.e., a genetically changed virus), the guards recognize it and are ready to destroy it.

Regardless of the methodology, a carefully calibrated recipe of ingredients goes into the vaccine preparation, including various additives to stabilize the biologic material, liquids such as sterile water to dilute it to the proper concentration, and then preservatives to keep the vaccine from degrading on its way to each recipient. Some vaccines also have adjuvants, ingredients that stimulate and boost the immune response. Adjuvants can be natural extracts, synthetic compounds, or various chemical particles, depending on the vaccine.

Vaccines can be produced in quantity through a variety of methods. The most common method for flu vaccines, for instance, employed for the better part of a century by now, is through egg-based manufacturing. The United States actually maintains secure, strategic populations of egg-laying chickens for this purpose. The candidate is injected into these fertilized chicken eggs and incubated for several days so the virus can replicate; then the fluid containing the virus is harvested, purified, and turned into the vaccine itself.

Another method is cell-based technology. In this approach, the viruses grow in lab-cultured mammal cells rather than eggs, with the virus-containing fluid then extracted and purified. It should be noted that no animals are harmed in this process, and it can be up and running faster than the cumbersome egg-based method.

Some vaccines, flu among them, can now be produced through synthetically re-creating a virus's genetic composition, called recombinant vaccines. The gene is combined with a baculovirus, one that infects invertebrates like insects but is not harmful to humans. It is then inserted into a "host" cell line, where it rapidly grows the viral antigen in bulk. Finally, it is collected, purified, and packaged.

Without getting into too much technical detail, there are six types of vaccines in use today. One method of producing a vaccine is to use the disease-causing virus after it's been killed but while it can still be recognized by the immune system. This is the concept behind the celebrated Jonas Salk polio vaccine. The hepatitis A vaccine is another example.

A second method is to use a live but attenuated (and therefore neutralized) version of the virus, as Albert Sabin employed for his subsequent polio vaccine, which was taken orally rather than by injection, making it easier to administer around the world. Another example is the MMR—measles, mumps, rubella—vaccine, which uses weakened forms of the viruses that can still replicate and trigger an immune response without causing the diseases themselves.

Today, in addition to inactivated vaccines like Salk's and live attenuated vaccines like Sabin's, there are four other routinely used technologies. They include subunit, toxoid, viral vector, and messenger RNA (mRNA) vaccines.

Subunit vaccines use one or more inactivated parts of the virus or bacterium to stimulate an immune response and are effective against hepatitis B, diphtheria, and shingles, for example. Toxoid vaccines, such as for tetanus, employ a controlled amount of the toxin produced by the bacteria. Viral vector vaccines use a harmless virus as the vector, or carrier, to deliver some of the virus's genetic material for the immune system to recognize and react to. Some of the new Ebola vaccines employ this methodology.

Scientists decide which type to use based on a number of factors. While these vaccines may differ, they have a common purpose: effective, durable immunity before one encounters the target pathogen.

The latest technology, messenger RNA, has been in development for the past two decades and was first employed beyond clinical trials against Covid. Once the mRNA is in the body, it "teaches" the relevant cells how to make a protein that triggers the immune response, producing antibodies when an individual is later exposed to the

actual virus. In the case of SARS-CoV-2, the mRNA approach recon-
structs the spike protein part of the virus, so the body temporarily
produces the protein, which allows the immune system to later recognize and respond to it.

One approach to respiratory viruses is a mucosal vaccine that is
sprayed in the nostrils rather than injected. The idea is to attack and
neutralize the virus in the nasopharyngeal area before it can make
its way to the lungs.

Within these six types of vaccines there are substantial differences in how they fulfill or handle the various aims of a vaccine: providing protection against infection and clinical illness; durability of
protection; route of administration (i.e., injection, nasal spray, or
ingestion); cold chain requirements for storage prior to administration; number of doses required with the initial series or boosters
after the first dose; safety; and cost. The unique complexities of each
vaccine's manufacturing process are also important as they determine the length of time it takes to make the vaccine as well as the
volume of vaccine that can be produced within months. Of all of
these characteristics of a vaccine, safety is the obvious priority.

Some vaccines provide excellent long-term protection. For example, the first dose of MMR vaccine is recommended for children
between twelve and fifteen months after birth. The second booster
dose is given between four and six years of age. This two-dose
approach results in 97 percent protection against measles well
through adulthood.[8] The measles virus has undergone very limited
genetic changes in the decades since the vaccine has been used,[9]
resulting in outstanding immune protection with the same vaccine.
The smallpox vaccine used prior to global smallpox eradication
resulted in a similar protection profile. In contrast, both influenza
and coronavirus vaccines must be updated frequently due to the constant genetic mutation of both viruses. A meeting of vaccine experts
is conducted by WHO twice a year to determine the influenza strains
targeted in the seasonal vaccine given annually. With even more
robust variant development in SARS-CoV-2 viruses, it can be argued
that Covid vaccines should be updated at least twice annually. And
even with annual influenza vaccination[10] and twice-yearly vaccination

for Covid,[11] protection against medically attended clinical illness is often only 30 to 60 percent. Think of measles as being on the "exceptional" end of the vaccine protection spectrum, with influenza and coronavirus at the "limited" end. It is important to note, though, that these limited-protection vaccines still save millions of lives around the world each year.

The holy grail of combating, or possibly even someday preventing, an influenza or coronavirus pandemic would be what's referred to as a universal vaccine for each one. There are different ideas and interpretations of how the term is defined, but the general concept is a vaccine that would provide robust and long-lasting protection against all forms or strains of the targeted virus. There is an entire body of research that is ongoing that is attempting to accomplish this lofty goal. It will require major new advances in both attacking the viruses with multiple components of the human immune system and discovering new ways to target the conserved parts of the viruses that undergo limited, if any, mutational changes over time. If universal influenza and coronavirus vaccines could be administered in childhood and result in immune protection for many years, as the measles vaccine does, we would have the potential to take the two major causes of devastating pandemics off the table. I firmly believe such an accomplishment is possible one day. And when it comes, it will be as consequential as the eradication of smallpox; maybe more so.

In the meantime, there is a second objective we are trying to bring to reality, known as game-changing vaccines. Within this category we further designate vaccines as "broadly protective" or "next generation." The first goal markers here are the next-generation influenza and coronavirus vaccines, which would represent an incremental improvement over the vaccines available today. They would employ a different strategy to elicit protective immune responses. The current influenza vaccines target the hemagglutinin head antigen on the virus, the part that often undergoes genetic changes. For the SARS-CoV-2 virus, the vaccines focus on the spike protein, which can also undergo rapid genetic changes. With game-changing vaccines, researchers are exploring ways to elicit a multifaceted immune response to parts of the viruses that do not undergo frequent changes, through which they are hoping to achieve improvement in durability, effectiveness, and breadth of protection.

The next goal in the search for substantial vaccine improvement is to realize more broadly protective vaccines. Recall how we explained earlier that influenza viruses are categorized by the hemagglutinin (H) and neuraminidase (N) proteins on their surface. The goal is to confer protection against all strains of a single influenza HA subtype, such as H1 through H18, multiple coronavirus spike proteins, or any targetable component of a virus that is conserved throughout all of its strains, forms, or variants. This research focuses on expanding the use of a wide range of investigational technologies that result in broader and more durable protection for both seasonal and pandemic influenza as well as expanding coronavirus vaccine protection to other betacoronaviruses, including the SARS-CoV (the cause of SARS) and MERS-CoV (the cause of MERS).

When we're faced with a potentially deadly virus with wings, the most important factors will be the speed of vaccine development and scaled-up availability, its effectiveness, and, as we've mentioned, its safety. Remember, even with the mRNA vaccine technology in hand at the start of the Covid pandemic, an estimated 70 million reported cases and 1.6 million deaths[12] occurred before the first vaccine (Pfizer-BioNTech) was approved for emergency use. Unless we have a near universal influenza vaccine or a pan-coronavirus vaccine already available, it will be as many as six months or more before an effective vaccine can be developed, tested, manufactured, and distributed widely. During that time, we will have to depend on therapeutic, anti-inflammatory, and antiviral drugs that are readily available and prove effective. They may not be able to prevent infection, but they should be able to put a dent in the number of serious or fatal cases.

One initiative may substantially change our vaccine development and availability timeline. I had the good fortune to be included in a series of meetings in Oslo and London in 2016 whose purpose was to envision whether a unique publicly and privately supported organization could accelerate vaccine research, development, and availability for pathogens of epidemic and pandemic potential.[13] Out of these meetings came CEPI, the Coalition for Epidemic Preparedness Innovations. CEPI is an international partnership of public, private, and philanthropic organizations launched in Davos, Switzerland, in 2017, whose mission is "to accelerate the development of vaccines and other biologic countermeasures against epidemic and pandemic threats so they can be accessible to all people in need."[14]

Vaccines can help prevent outbreaks from becoming humanitarian crises, and CEPI supports not only the development of vaccines but also platform technologies and manufacturing innovations to combat priority diseases for which no licensed vaccines are currently available. The organization leverages its unique position within the global health and R&D ecosystems to address market failures, building on its track record of bringing together public-sector, private-sector, and academic partners. By late 2024, it had been able to enlist more than thirty investing organizations,[15] garnering $3.1 billion in support.[16]

CEPI also engaged more than 180 partners[17]—including CIDRAP—by 2024, researching and developing vaccine and biologics for seven infectious diseases. The organization's website explains, "CEPI's priority pathogens include Chikungunya, COVID-19, Ebola, Lassa Fever, MERS, Nipah, Rift Valley Fever, and novel viral threats with epidemic or pandemic potential (also known as 'Disease X')."[18]

While I strongly support the remarkable CEPI leadership team, led by chief executive officer Richard Hatchett, and endorse the organization's overall efforts, I am concerned that the priority pathogens initially chosen for primary investments did not include influenza, one of the two virus families I believe most likely to cause a pandemic. Fortunately, it does include the other: the broad family of coronaviruses. To achieve a potential game-changing influenza vaccine will take major resource investment, far beyond what government and philanthropic organizations are currently investing. Broadly protective coronavirus and influenza vaccines offer the best hope of a highly effective response to the next pandemic. CEPI is now working on the coronavirus challenge and exploring the inclusion of influenza as a Disease X prototype. This is a potential game changer for achieving broadly protective vaccines for the highest-risk viruses that will likely cause the Big One.

The absolute best-case scenario whenever such a pandemic emerges would be to have these game-changing vaccines already developed and available, and for at least part of the population to have been vaccinated. CIDRAP led more than a hundred international vaccine experts in developing "roadmaps" for each of these efforts, which, if initiated, would take years and require government investments of many billions of dollars. But as the staggering costs of the Covid pandemic, as well as the 1918 Great Influenza, have demonstrated, whatever the investment figures, they will be insignificant compared with the human and financial price that will *not* have to be paid when the Big One or any other airborne-transmitted viral pandemic occurs. As Dr. Seth Berkley, the highly regarded former CEO of Gavi, puts it, "Any investment into effective prevention or mitigation of an airborne viral threat is money well spent compared to the eventual cost."[19]

The fact that vaccines against SARS-CoV-2 were developed in record time on the shoulders of decades of research into mRNA technology and methods should give us encouragement *and* circumspection. Encouragingly, other countries delivered vaccines using various technologies relatively quickly, too. In addition to mRNA vaccines, ten other vaccines using recombinant and inactivated platforms were produced in Belgium, China, India, Russia, Korea, the United Kingdom, and the United States.[20] The protection offered by many of these vaccines against serious illness and death within weeks of being vaccinated was nothing less than spectacular. But the time frame in which immunity waned—a matter of months[21]—was deeply disappointing, requiring us to chase our own tails in re-administering booster doses. And the waning immunity after so much initial promise only added to the vaccine resistance and skepticism that became one of the prime political hot buttons during the Covid pandemic.

Dr. Florian Krammer is one of the brightest vaccine researchers in the world. Trained in his native Austria, Florian is professor of vaccinology at the Icahn School of Medicine at Mount Sinai Hospital in New York, as well as holding other prestigious positions. When he talks about what is needed to prepare for an airborne viral pandemic, smart people listen.

"There are two aspects [to vaccine development for the next pandemic]," Florian says. "The first is mRNA vaccines and their potential, and the second is what [else], in general, should be done. First, mRNA is a great development, very powerful, very quick.... And the immune responses we've seen are pretty amazing, albeit waning, from the initial antibody peak to the stabilization stage.... But there are issues: mRNA is one part of the arsenal that we have against viruses. There are issues in terms of [its] not working for all viruses.... [For Covid-19], efforts were good at making vaccine that was protective against severe disease but not good in producing a vaccine that prevents infection. I'd put money into designing a vaccine that protects against infection and onward transmission of the virus."[22]

Again, without getting more technical than we need to be, Florian explains there are two types of responses triggered by the newest vaccine platform and the challenge of an evolving virus. "The mRNA vaccines are especially good at driving antibody response: a certain population of B cells called plasmablasts. The body produces them as the initial cavalry. Those cells only live for about ten days and then they're gone, and antibody levels start to drop.... Then at some point the levels stabilize, triggered by another type of B cell that has had time to mature and make antibodies in the bone marrow. Once you have them, they don't go away for a while.... If the virus didn't change, most of us would still have good protection from infection. [But] if you only had the ancestral [original] Covid vaccine...you'd have little protection, not because the vaccine is weak, but because it's a different virus."[23]

The current coronavirus vaccines have proven to be good for individual protection against serious disease or death for a limited period, though not very good for preventing infection and transmission. The yearly flu vaccines are, at best, fair, offering limited protection against serious consequences that starts to wane after several weeks, so it becomes necessary to strategize when, as the flu season approaches, to get vaccinated. Eventually, any discussion of truly effective vaccines for either of these airborne pathogens comes down to the universal vaccine that many of us in public health consider to be a top priority.

Is mRNA the best platform? Many experts aren't sure, despite the experience with SARS-CoV-2. "mRNA is everyone's darling," says Dr. Bruce Gellin, former senior vice president and chief of global public health strategy for the Rockefeller Foundation's pandemic prevention initiative, former deputy assistant secretary for health and director of the National Vaccine Program Office. "But it's not the solution for all vaccinations. If you bet the farm on mRNA, at some point it won't be the solution you need. Redundancy and diversity will be the key."[24] This is one thing we should credit Operation Warp Speed for. It didn't put all its vaccine eggs in one basket, instead supporting several platforms to see which worked best and could be developed fastest.

Richard Hatchett, CEPI CEO, thinks that mRNA-based vaccines may turn out to be a good short-term or first-line vaccine defense because their flexibility allows them to be developed relatively quickly, even if their immunity protection fades over a period of months.[25] If that turns out to be the case with the next pandemic, the mRNA protection would still buy time to develop a vaccine with more lasting protection.

Admittedly, all the challenges to creating more effective and long-lasting vaccines are daunting. "With Covid, if we find the epitope [the part of the viral antigen that binds to a specific antigen receptor on the B cell], that should be a good target," Florian says. "At least, that was the initial idea. But that didn't take into account what would happen if you put pressure on the epitope. That made the [SARS-CoV-2] virus change, and that made it more complicated."[26]

What we did at CIDRAP on the critical influenza front is what we'd like to see as a model for ongoing and future influenza and coronavirus vaccine research and development. I believe it is the most important step we can take to prepare for the future Big One.

ROADMAPS

The 2009 H1N1 influenza pandemic was our last live-fire test of the world's pandemic vaccine response capacity before Covid. Despite our preexisting seasonal influenza programs, developing a pandemic vaccine, getting it approved by the FDA and other countries' regulatory agencies, and manufacturing it in sufficient quantity for the 6.8 billion world population was a challenge that was largely unmet, even in high-income countries. The United States was one of the first nations to roll out pandemic influenza vaccines — specifically, the fifth country to start vaccinating for H1N1, just twelve days after the first two, China and Oman, began their campaigns.[27] Still, the first shots didn't get into people's arms until more than six months after the first human H1N1 cases appeared in Mexico[28] — and this was using old technology to produce vaccine, like one for regular seasonal influenza. The vaccine was not widely available here until after the second, largest peak of cases, which occurred from late August through late November.[29] For much of the world, significant vaccine shortages persisted for the duration of the pandemic.[30]

As challenging as 2009 was, subsequent reviews of the overall pandemic response by governments and public health agencies did not result in any substantial new findings or recommendations about how to improve on the vaccine response for a future influenza pandemic. I found it particularly concerning that the discussions about our vaccine response were largely focused on when the vaccine became available, with little attention paid to how well the vaccine protected. At the time, the prevailing belief was that seasonal flu vaccines were 70 to 90 percent effective in protecting against infection,[31] but a study of the 2009 pandemic vaccine conducted by the CDC and local researchers in four US communities found it provided significantly less protection.[32]

This should have been a major wake-up call. We needed to understand how well influenza vaccines really protect us and what we can do to substantially improve them. In late 2010, I invited colleagues from the Marshfield Clinic in Wisconsin and Johns Hopkins Bloomberg School of Public Health to CIDRAP for a deep-dive review of all the studies evaluating the effectiveness of influenza vaccines. We eventually zeroed in on thirty-one studies, conducted between 1967 and February 2011, finding an average level of protection against illness requiring medical care at 59 percent for adults eighteen to sixty-five years of age[33]—a stark contrast to what the public health and medical communities were promoting at the time.

Over the next eighteen months, our research group submitted a manuscript titled "The Compelling Need for Game-Changing Influenza Vaccines: An Analysis of the Influenza Vaccine Enterprise and Recommendations for the Future," summarizing our findings to four medical journals. All rejected it, with reviewers citing the generally accepted 70 to 90 percent efficacy. Finally, in late October 2011, the journal *Lancet Infectious Diseases* published the paper, and a year later, CIDRAP published an exhaustive 160-page document. We argued that a major barrier to the development of game-changing influenza vaccines was the misperception that the available vaccines were highly effective.

In January 2018, Mark and I authored an opinion piece in the *New York Times* entitled "We're Not Ready for a Flu Pandemic."[34] Unfortunately, the piece could just as easily have been written in 2024. Nothing we can do to prepare for the next influenza pandemic could save more lives and minimize the social, political, and economic impacts than having universal or near-universal vaccines readily available worldwide that offer high levels of protection for years after one or two doses. Think: What if we could vaccinate the world *before* the Big One emerges and stop a frightening virus in its tracks?

But back to current reality.

Eventually, public health agencies recognized the underperformance of influenza vaccines and began efforts to address the challenge. In particular, the CDC and the NIH began to support influenza vaccine effectiveness studies and research efforts to find more protective vaccines, and more reliable and timely ways to manufacture them.

In 2019, the Sabin-Aspen Vaccine Science & Policy Group found that "fragmentation and lack of goal-oriented coordination"[35] were the primary constraints on progress toward new influenza vaccines. Overcoming these challenges requires unprecedented collaboration through new, innovative approaches addressing unresolved scientific issues and the high costs of basic and translational (i.e., connecting different areas of investigation) research. To provide clear and compelling guidance for this work, CIDRAP developed the Influenza Vaccines Research and Development Roadmap (IVR).[36]

The concept came from the Global Funders Consortium for Universal Influenza Vaccine Development, initially convened in 2017.[37] With a ten-year vision, the purpose was to improve the effectiveness and production of strain-specific seasonal influenza vaccines and advance research and collaboration in the development, licensure, and manufacture of durable, broadly protective, or universal influenza vaccines to benefit all countries.

In response to this request from the consortium, in 2019, the British Wellcome Trust provided funding to CIDRAP to coordinate development of the roadmap. Over two years, CIDRAP developed the IVR in partnership with a steering group from CIDRAP, WHO, the Bill & Melinda Gates Foundation, the Sabin Vaccine Institute, the Wellcome Trust, the CDC, and the Task Force for Global Health, as well as an internationally and professionally diverse seventeen-member IVR task force of global subject matter experts and leaders in the field. The group identified six topic areas and fifty-seven R&D milestones within those areas. The roadmap provides a blueprint for action that won't produce new vaccines in weeks or months but rather years. Obviously, we can't wait for the next pandemic to begin this process. And we will only accomplish our goal if we adequately fund research efforts.

The NIH is a global leader in the kind of influenza vaccine research detailed in the roadmap. In September 2019, the National Institute of Allergy and Infectious Diseases announced the launch of the Collaborative Influenza Vaccine Innovation Centers (CIVICs) program. The seven research centers in the CIVICs network work together to develop more durable, broadly protective, and longer-lasting influenza vaccines. As critical as this research is in developing more effective vaccines for both pandemic and seasonal flu, in the five fiscal years beginning October 2019, it has been funded at only $46 million to $64 million per year.[38] That financial support is a mere rounding error for federal departments like the Department of Defense or Department of Health and Human Services. This program should be funded at much higher levels. The potential payback could be immeasurable.

In planning and preparing, what applies to flu applies equally to coronavirus. So, in 2022, with support from the Rockefeller and the Bill & Melinda Gates Foundations, and through the collaborative efforts of fifty international experts, CIDRAP created a comprehensive Coronavirus Vaccines R&D Roadmap (CVR) for developing broadly protective coronavirus vaccines. Beginning in 2024, CEPI began supporting the roadmap effort. The CVR, like the IVR, provides a framework and six-year timeline for essential research, leadership, and investments, while offering a detailed assessment of specific scientific, regulatory, and logistical challenges. Many efforts are already underway, and the CVR offers the structure and time milestones to ensure this work is well-aligned and focused on building the coordination, leadership, and investment essential for achieving the roadmap's ambitious objectives: to accelerate the development of durable, broadly protective coronavirus vaccines that reduce severe illness and death (and potentially prevent infection) due to current and future coronaviruses known to infect humans, and those at risk of spilling over from animals; to mitigate the impact of future coronavirus epidemics worldwide; and to provide solutions that are suitable for use in all regions of the globe, including remote areas and low- and middle-income countries.

In May 2023, the Biden administration announced a $5 billion Department of Health and Human Services project called "Next-Gen," a major investment in the rapid development of next-generation coronavirus vaccines and treatments.[39] You might expect me to be highly supportive of such an effort, but I have serious reservations about this approach. There is a prevailing mindset among some of my colleagues and many in the political and policy worlds that mRNA vaccines were a ten-month miracle from first concept to completing clinical trials. But they forget that mRNA technology was under development for several decades before it was used for a coronavirus vaccine, with immune-evasive variants still remaining a significant challenge. Additionally, a single-minded focus on coronavirus is far too limited to provide the solutions we need in preparing to face pandemics; and a $5 billion onetime government investment—a tiny fraction of what the United States devotes to weapon systems procurement—is only a minimal down payment on the research and development needed to accomplish this critical goal. Consider that if any airborne virus begins to spread in the human population and sparks a pandemic with a fatality rate even 3 to 5 percentage points higher than Covid, the world will be going to war against a terrifying microbial enemy. It would be far more deadly than any pandemic in living memory, or any military conflict since World War II.

I am not alone in my concerns. A June 26, 2023, article in the *New York Times* began, "Efforts to develop the next generation of Covid vaccines are running up against bureaucratic hassles and regulatory uncertainty, scientists say, obstacles that could make it harder to curb the spread of the coronavirus and arm the United States against future pandemics."[40]

I can't overstate how the "bureaucratic hassles and regulatory uncertainty" serve as showstoppers for even the best, and best-funded, vaccine research and development programs. One of the primary reasons Operation Warp Speed accomplished what it did in getting vaccines through to FDA Emergency Use Authorization (EUA) approval was an agreement among all the federal agencies involved to streamline the hassles and uncertainty at a level that I had never witnessed before in my career. But now we are back to the pre-pandemic vaccine approval mindset. We must never compromise safety and demonstrated vaccine effectiveness, but at the same time, we need a revolution in how we support vaccine development and eventual licensure. I don't believe NextGen addresses this critical issue.

NextGen also presents a special challenge for bringing in the scientific and financial support of the pharmaceutical companies. The larger companies have little to no incentive or interest in developing new vaccines, since, in their judgment, it may be decades before they are needed for the next pandemic. Private investors, the primary financial lifeline for small biotechnology companies, are largely sitting out as of this writing. NextGen is the only real hope for these small companies surviving in the expensive world of vaccine research and development.

I was seriously distressed by several off-the-record conversations I had in the early summer of 2023 with senior officials in the White House and HHS, where there was unbridled optimism that NextGen would produce a new and better vaccine technology by the fall of 2024. I knew there was not a snowball's chance in hell that was going to happen. And sure enough, it didn't. No company, regardless of size or scientific expertise, is likely to achieve new, game-changing technology in eighteen months. As our roadmap process has clearly documented, even with some of the best minds in coronaviruses, immunology, vaccinology, and policy and financing, our research agenda will take time, with each step building on what we learn. But now we have created an expectation that the $5 billion program will fix all our vaccine challenges and we don't need further investment. So, if NextGen funds dry up after that, what will we have accomplished? I can imagine legislators saying to the health and medical establishment, "We gave you $5 billion, and you didn't come up with game-changing results. Why should we give you more money and a longer runway?"

Yet imagine what this level of financial support over five to eight years could do if it were invested in a program similar to the NIH's CIVICs program for influenza. In reality, we may not have six or ten years before another 1918-like influenza pandemic occurs, so we must work as quickly as we can on these vaccines because they represent the one realistic "Get Out of Jail Free" card with the next pandemic. But if we oversell what we can do in short order, we jeopardize both our long-term results and our credibility.

SUPPLY AND DEMAND

Dr. Bruce Gellin thinks a lot about what the early days of a new pandemic would look like. "The whole problem is about speed and volume," he says. "Everyone talks about time to the first dose, but no one talks about time to the last dose [i.e., when everyone has been vaccinated]. In between is balancing out what supply is available and knowing what will ship over time. Unless someone's going to mothball all the equipment and personnel needed, you'll always have a problem up front.... So, part of the thinking needs to be an acknowledgment of the mismatch between supply and demand at the beginning." Then, Bruce notes, "the scenario plays out on both the country and the community level. How do you balance that? From a production standpoint, you need to acknowledge where that's going to happen.... Frankly, that was the problem with Operation Warp Speed. It was all about vaccine development, and barely about getting [distribution] up and running.... They're two separate worlds that have to connect."[41]

COVAX was an alliance established by Gavi, CEPI, WHO, and UNICEF to provide equitable distribution of SARS-CoV-2 vaccine around the world. It was a noble effort, getting out millions of doses in low-income countries, but it fell far short of its goal. Partly, this had to do with the inability to build local manufacturing capacity and develop expertise in a limited period of time. Reluctance to share pharmaceutical intellectual property with other countries to manufacture the vaccines was not the major issue it was made out to be. Some of the problem had to do with the upper-income nations that produced most of the vaccine supply initially hoarding it for their own populations. Ultimately, the coalition couldn't control the supply. Ironically, by the third year of Covid, Bruce notes, "There were 2.5 billion doses that governors were trying to figure out what to do with."[42] We often hear that with a deadly, airborne infectious agent, no one is completely safe until everyone is safe. While this statement might be a bit hyperbolic, the sentiment behind it is genuine, and we fell far short during Covid, when much of the world remained vulnerable as many in the upper-income nations chose to refuse vaccination. Bruce warns that next time, we will have to do better.

Seth Berkley says, "We will need a global stockpile" that doesn't depend on one country's politics.[43]

Bruce adds, "With local manufacturing, there's a reasonable effort underway to think that through. But it needs to be thought through not just during a pandemic, but sustainably. All things are in play: having technology, technicians, the regulatory system, and having a business model that works over time."[44]

Sustainability of such overseas manufacturing facilities is the aspect many of my expert colleagues consider a critical challenge but one that is crucial for future pandemic preparedness. Christopher "Chris" Chadwick was a technical officer at WHO in Geneva before going to the CDC. He has worked on the effort to spread viral drug and vaccine manufacturing capacity around the world. "In 2021, the WHO launched the mRNA Vaccine Technology Transfer Hub in South Africa," he says, "charged with developing technology and doing R&D for Covid [and influenza] vaccine that would transfer to fifteen manufacturers in low- and middle-income countries."[45]

Chris says that WHO has a working group assessing business models. The model that seems to make the most sense is to have facilities capable of producing vaccines that in non-pandemic times (i.e., most of the time) would be producing seasonal flu vaccine and biologics that would be marketed to other low- and middle-income countries, to help make the manufacturing facility sustainable.

One factor standing in the way, Chris says, is that there is little demand for seasonal flu vaccines in low- and middle-income countries. "We don't know that it'll be technically successful. We're hopeful, but that's part of the model—that the mRNA is the test case.... We have a training hub to make sure a skilled workforce is being trained in biomanufacturing. And in that way, we're not completely tied to mRNA vaccine technology."[46]

If such an initiative works, not only will it be sustainable for when the next pandemic arrives, but it will also be flexible enough to produce whatever is needed at any given time, such as polio vaccine.

The fifteen manufacturers in the WHO project run the gamut from mature and experienced facilities in places like Brazil and Argentina to nascent start-ups in Africa.[47] The idea would be that they could all be tasked with producing according to the same vaccine formula. CEPI is trying to universalize the approval process, which would mean that each country producing vaccine would not have to go through its own clinical trials.

"It's a long process to set up manufacturing capabilities," Chris concedes, "and even longer to achieve sustainability. It's a huge investment, but some capacities you can continue in a pandemic period with lower resources from the national government. But with the intricacies and complexities of vaccine production, it all has to be stable."[48]

While the focus on manufacturing mRNA vaccines is critical in preparing for the next coronavirus pandemic, it very well could be that the next pandemic is the result of a new influenza virus. Erin Sparrow and colleagues at WHO and in Australia published a sobering paper in 2020 describing the global manufacturing capacity in 2019 for influenza vaccine. They found the estimated annual worldwide seasonal influenza capacity to be around 1.48 billion doses, with potential production capacity of up to 4.15 billion doses.[49] Since two doses of vaccine are required to be fully vaccinated, a maximum of 2 billion people could be vaccinated in the first year of the next influenza pandemic.

The same study identified eight assumptions that will impact vaccine manufacturing capacity and determine the actual production. These include that manufacturers would not be in the midst of producing seasonal influenza vaccine (since it would be difficult to impossible to switch production quickly); that it could be grown in eggs as well as cell cultures to increase capacity and that egg-laying poultry would not be compromised by an avian viral strain; that there would be sufficient capacity for manufacturers to get the vaccine into syringes; that there would be adequate protection for workers so that illness would not affect production; and that there would be enough reagents and adjuvants (as well as basic supplies of vials, syringes, and the like) to produce and deliver vaccine. The absence of some of these factors could be deal breakers. As it is, we have a complicated and antiquated system to produce vaccines at scale, and even to achieve what the study outlined, a lot of planets would have to align for all of these conditions to be met.

It's simplistic to think we will have solved our vaccine challenge if we identify an effective and safe vaccine. We also must consider the manufacturing requirements, speed of production, cold chain storage needs, dosing, delivery methods, and expert staff required for manufacturing, in particular in low- and middle-income countries.

What we are describing is not what has generally been done in past epidemic and pandemic planning and preparation, which tends to be reactive rather than proactive, getting underway only when the threat is already in front of us. We have to acknowledge and accept that no pharmaceutical company or academic center is going to provide the substantial financial resources to build and maintain the vaccine-manufacturing infrastructure we have described here. Government is the only entity that can make this happen. There will be substantial pushback by many governments against helping support vaccine-manufacturing capacity for other countries. But as we know, pandemics are global, and to protect ourselves, we need to provide that same protection for the world.

To have any real hope that we can produce effective vaccines in the earliest days of the Big One and distribute them quickly to cover all global demands, we need to adopt the military model of procurement. The DOD does not wait until war breaks out to design and build an aircraft carrier. With the decade or more required to specify, bid out, design, build, fit out, and undergo sea trials, an aircraft carrier that wasn't battle ready at the start would not be much use. The same is true for our war against pathogens. The aircraft carrier, or fighter jet, or whatever weapon we're talking about, is built by private industry and paid for by the government with guarantees up to completion. This is the example we should follow for both antiviral agents and vaccines. Unlike with an aircraft carrier, there is opportunity here for the private sector—the pharmaceutical industry— to make a profit selling its products. But for the business model to work and the company to devote the necessary resources, agreed-upon purchase levels must be guaranteed, whether or not the pandemic occurs in the immediate future. We call this contingent financing, and it must be legally binding, which would require a strong political commitment, not only in the United States but in other high- and middle-income countries. This is the only way that real preparation happens.

FROM VACCINE TO VACCINATION

Finally, we can have highly effective and safe vaccines readily available in the early days of the next pandemic and still have a largely unprotected public. That's because we fail to turn vaccines into vaccinations. A major part of this effort involves effective communication and managing public expectation, which we will address in the next chapter. But to make progress in this area, we must support major research efforts into understanding the psychology and sociology of vaccination uptake in a pandemic setting among all members of our society. Public health became its own worst enemy regarding the public's confidence in the Covid vaccine when the very early vaccine trial results reported that the vaccines were at least 90 percent effective[50]—"Now you can take off your masks"—but then showed, six months after vaccination, less than 20 percent protection against illness and infectiousness.[51] In summer 2021, the message to the fully vaccinated was "Put your mask back on."

A critical lesson is that when multiple vaccinations are required over time to sustain immune protection, there will be a substantial decline in the levels of up-to-date vaccinated people. In the six months following September 2022, when the bivalent mRNA booster dose became available, only 42 percent of US residents sixty-five years of age or older received a booster dose,[52] despite these older residents being at increased risk for serious illness, hospitalization, and death.[53] When a pandemic lasts two or three years, expect a major drop-off in rates of vaccination over time when vaccines need to be administered every six to nine months.

Highly effective and durable vaccines available very early in an emerging pandemic will change the course of history if accepted by the public. If the Big One is upon us and we have not changed these factors, we can expect illness, death, and economic disruption far beyond what we witnessed with the Covid pandemic.

THERAPEUTICS

Vaccines are designed, developed, and delivered to prevent someone from becoming infected. But when vaccines don't exist for a given infection, or the number of vaccine doses is inadequate to meet demand, or vaccination reduces the risk of serious illness and death but doesn't prevent it, therapeutics become a critically important part of the medical and public health response. If vaccines are still being developed in the early days of the next pandemic, we will need so-called small-molecule drugs with broad-spectrum activity against influenza and coronaviruses. We refer to them as "small molecule" because unlike, say, anticancer agents, which are often made up of huge molecules, small-molecule drugs can easily slip into cells and are designed to prevent viruses from taking over the cell's reproductive mechanism and churning out more copies of themselves. There are various ways these drugs can work. For example, protease is one type of chemical necessary for many types of viruses to reproduce, so some antivirals are protease inhibitors. Today we have approved drugs for Covid, including Paxlovid (nirmatrelvir/ritonavir), Veklury (remdesivir), and Lagevrio (molnupiravir).[54] For influenza, we have Tamiflu (oseltamivir), Relenza (zanamivir), Rapivab (peramivir), and Xofluza (baloxavir marboxil).[55]

Just as with vaccines, I believe our therapeutic-preparedness focus should be twofold: First, stockpile the current approved drugs we have for influenza and coronaviruses at volumes adequate to treat 25 to 40 percent of the population; and second, research and develop new broad-spectrum drugs with the potential to be safe and effective against all influenza A viruses—the ones that cause pandemics—and novel coronaviruses that might infect humans. This doesn't mean we neglect investment in drugs to treat infections we consider of significant epidemic potential, such as Ebola, Lassa fever, Nipah, and Marburg. But we must prioritize drugs that can quickly be distributed and used during the earliest days of a pandemic.

With the drugs approved for Covid and influenza, we have a potential first line of defense against whichever virus causes the Big One. However, we need a comprehensive study to determine more precisely how effective these drugs might be against a range of influenza A viruses, including those of the eighteen hemagglutinin and eleven neuraminidase subtypes that may cause a human pandemic. Drug treatment effectiveness data are available for the seasonal influenza subtypes. For influenza viruses that so far have not become part of the seasonal flu picture, studies will need to be laboratory-based estimates (in vitro) of how well the drugs impact the virus in the test tube. We don't want to wait for a pandemic outbreak to find out! The same is true for coronaviruses. These studies can provide important information for determining which of the current drugs for seasonal influenza and Covid will be effective against the virus that causes the next pandemic.

Dr. Stanley Perlman, professor of microbiology and immunology at the University of Iowa's Carver College of Medicine, is one of the leading experts on the pathogenesis of coronaviruses. He acknowledges how difficult it is to create a new vaccine, particularly early in a pandemic, when it could have the most impact on spread. "Vaccines are hard," he says, "which is why I like the antiviral idea."[56] And he wants extensive research to begin now.

We need an international stockpile of drugs that, as we have recommended for N95 respirators, can be updated and rotated out to ensure their stability when needed. We know that oseltamivir, the anti-influenza agent marketed as Tamiflu, turns to powder over time. So we have to be prepared to actively manage a stockpile for years on end. If we are serious about having the potential to greatly reduce the morbidity and mortality of the Big One, some of the stockpiled drugs will become outdated, because even with major efforts to use these drugs to treat cases occurring seasonally (i.e., influenza) or with ongoing transmission (i.e., coronavirus), the turnover may not be sufficient to rotate the quantity of drugs we need for an international stockpile system. This is the price we must pay for preparedness. We can't be in the midst of the Big One and suddenly find that the antivirals we have stockpiled have outlived their effectiveness.

Realistically, we must also recognize the possibility that the new virus causing the Big One will not be susceptible to the drugs in our stockpile. It's a gamble, but maintaining an active, up-to-date, and ready supply of antivirals with known effectiveness will increase the odds in our favor.

Seth Berkley has stressed the necessity of having manufacturing and distribution facilities throughout the world already in place when the need arises. The global demand for small-molecule drugs to combat influenza and coronavirus, and the diagnostics to rapidly confirm infection, will require a new international apparatus with adequate resources and the stockpiling expertise to ensure that high-, middle-, and low-income countries all have access to the necessary medical countermeasures in the early days of the pandemic up through the arrival of vaccines. Mark and I believe a group like the Gavi management team would be well positioned to manage this.

We have to face the fact that there is no compelling financial motivation for pharmaceutical companies to devote their limited space and resources to conduct basic research and development and produce antivirals. Pharmaceutical companies large and small are exiting the infectious disease drug and vaccine markets because that is not where the greatest profit potential exists. In 2024, pharmaceutical market experts estimated that by 2031, the infectious disease drug and vaccine market worldwide would be approximately $184 billion,[57] and in 2030, the oncology market will be $526 billion.[58]

That's why it was not a shock in the summer of 2023 when Janssen, one of the leading global infectious disease pharmaceutical companies, announced it was largely exiting the infectious disease market.[59] Shortly after the announcement, the head of the Japanese pharmaceutical company Shionogi called on G7 governments to lead on fixing the market for infectious diseases or run the risk that other companies would leave the market.[60] In recent months, I learned of two pharmaceutical companies that quietly shelved their promising research on coronavirus antivirals because there wasn't enough money in it to justify the opportunity costs and commitment to stockholders. It will largely be up to the governments of the world to step up and provide the investment resources.

Recent initiatives may provide some nongovernmental support for new drug development. In 2022, the Novo Nordisk Foundation, the Bill & Melinda Gates Foundation, and Open Philanthropy announced their initial coordinated commitment of up to $90 million to support the Pandemic Antiviral Discovery (PAD) initiative, "to catalyze discovery and early development of antiviral medicines for future pandemics."[61] Influenza and coronavirus infections, and the family of viruses that includes measles, are noted in their publicly stated priority areas. In 2024, the initiative provided $20 million in funding for eleven research projects worldwide pursuing research and development of novel antiviral drugs for pandemic influenza.[62] While this effort is a positive step forward, like other initiatives we've described, it will require many times this amount to realize effective influenza and coronavirus MCMs.

In March 2023, a nonprofit consortium of seven pharmaceutical companies established the INTREPID Alliance, with a goal of accelerating discovery and development of antiviral drug candidates that would be ready for clinical trials when a future pandemic emerges. The alliance is generating listings of antivirals that could be quickly deployed into Phase 2/3 clinical trials following an emerging pandemic. I believe that the alliance is a critically needed partner in the antiviral drug research and development environment, much like CEPI has become to vaccine research and development. I can only hope that the alliance is able to achieve similar status, securing funding and public and private partners to move antiviral drug discovery forward with urgency. It could make the critical difference in having lifesaving drugs for treatment months before vaccines are available.

The next basic challenge will be getting doctors to prescribe drugs and getting patients to take them. Our track record in this area, too, has been extremely disappointing.

Overall, Paxlovid use in the United States has been much lower than initially expected after the drug was approved in December 2021. As of January 2023, only 6.7 of 10 million available doses had been used.[63] Yet, during this thirteen-month period, 276,379 Covid deaths were reported in the United States,[64] despite abundant evidence that Paxlovid can significantly lower the risk of serious illness, hospitalizations, and death for seniors and those with compromised health.[65] It has also been shown that Paxlovid significantly lowers the risk of developing long Covid.[66] The perceived risk of rebound Covid, among both patients and healthcare providers, has likely contributed to this low uptake. Yet multiple studies have found that the rate of rebound Covid illness is similar in those who have not taken Paxlovid.[67]

Two other tactical weapons against new pathogens of pandemic potential are anti-inflammatories and monoclonal antibodies. Viral infection causes inflammation in the body, which is actually an immune system reaction. But the immune system can overreact, which makes the accompanying inflammatory response a problem. On a basic level, cytokines, a critical component of the immune system, are small proteins that "alert" the relevant white blood cells to rush to the location of the infection to fight the invaders. In a cytokine storm—an immune overreaction—the continual feedback loop between cytokines and the defensive cells can clog airways and inflame tissues, and ultimately cause organ shutdown. Steroids, which fight inflammation, are one part of the solution.

Monoclonal antibodies are synthetically produced proteins designed to stick to a specific part of the antigen, such as the spike protein of SARS-CoV-2. They act similarly to the body's own immune defenses but do not have to "wait" for an immune response to be mounted. But they are time-consuming and expensive to produce, and one monoclonal will not necessarily work for a different variant of the virus. So, while an often-useful weapon, these have not proven a panacea thus far.

Many friends and colleagues, as well as hundreds of people, including seniors and those with immune-compromised conditions, reached out to me from around the country after being told by their physician or health plan triage nurse they didn't need to take Paxlovid or another antiviral even when their underlying health status put them in the risk category for serious illness. I intervened once when the triage line staff at a large VA hospital told all newly diagnosed Covid patients they were to get Paxlovid only if they were seriously ill and hospitalized. This had gone on for at least six months before my follow-up with the head of the infectious disease group at the hospital corrected the triage line procedure. I witnessed many examples of physicians being poorly informed about the appropriate and timely use of Paxlovid or other Covid antiviral. A major educational effort needs to be undertaken before the next pandemic to ensure that the best in evidence-based medicine is followed.

Along with the overall low uptake of Paxlovid, distribution has not been equitable. According to a CDC report published in October 2022, data from nearly 700,000 people seeking Covid treatment across thirty US sites from April to July 2022 showed that Black people were 36 percent less likely and Hispanic people were 30 percent less likely to be prescribed Paxlovid than white people.[68] Equity issues with Paxlovid distribution have also been seen on an international level. As of May 2023, high-income countries that make up 16 percent of the world's population had received more than 70 percent of Paxlovid doses.[69]

We have lots of work to do before the next pandemic to find new and more effective drugs, make them widely available, and educate the medical and public health communities, as well as the general public, as to their importance in reducing severe illness, hospitalizations, and deaths.

During the Biden administration, Congress established the White House Office of Pandemic Preparedness and Response Policy, under the direction of retired Major General Paul Friedrichs, MD. This was an encouraging development but nowhere near the level required to manage the Big One. Imagine if the chairman of the Joint Chiefs of Staff had just a few assistants to implement the military's command and control structure, and the DOD lacked control of its overall budget. This is the reality of the federal government's infectious disease response, spread over a multitude of departments, all with their own budgets and priorities. While General Friedrichs made meaningful important progress coordinating these agencies, much remains to be done.

DIAGNOSTICS

From the first days of the Covid pandemic, the ability to determine who was and who was not infected with SARS-CoV-2 was a top medical and public health challenge. Knowing one's infection status was critical to the individual in seeking medical care and for reducing contact with others to limit transmission of the virus. It was also essential from a public health perspective, as the epidemiology of Covid could only be determined by the case data collected at the community level. Accurate data on who was and who wasn't infected, who was ill and infected, and who was infected but not ill, enabled us to define patterns of transmission. In turn, this allowed public health officials to make recommendations on ways to reduce one's risk of becoming infected and, in some situations, even predicted what the community burden of illness might look like in several weeks. And it armed healthcare systems with the information they needed to anticipate their resource needs to treat and care for Covid patients in the weeks to come.

Most of us know the story of the CDC's unfortunate release of a Covid PCR test on February 4, 2020, that turned out to be highly unreliable.[70] This started a cascade of more significant failures in the Covid testing system that have not been properly addressed and that serve as potential points of failure for the Big One.

First, we never established what we needed in a Covid test result to serve all purposes. And because the commercial and government testing capacity was so unprepared for the overwhelming population-based testing needs, the FDA made decisions to approve tests for public use via the EUA process that would never have been approved pre-pandemic. Nor would the US government have provided billions of dollars of support to many companies, some of them start-ups that may just have been trying to cash in on the market for pandemic testing. Home tests used lateral flow technology, a simple and rapid assay designed to detect the presence of a specified substance in a body fluid sample. Home pregnancy tests, for example, use lateral flow techniques to identify a particular hormone in a subject's urine. But the Covid tests lacked sensitivity to detect infections in the first few days. I was honestly horrified when the FDA approved the Abbott BinaxNOW lateral flow test, which became a widely used home diagnostic test. For EUA approval, Abbott submitted test results from 460 samples from symptomatic patients. One hundred and seventeen were PCR positive (using a laboratory-based test with higher sensitivity), yet only ninety-nine of those were BinaxNOW positive. This means that 15.4 percent of true positives were missed by the BinaxNOW test. And since 343 of the 460 were not Covid-positive by PCR, that means the FDA approval for BinaxNOW was based on testing only 117 positive samples.[71]

The conclusion from these data, together with numerous other studies, is that the home testing kits using lateral flow technology produced lots of false-negative results, particularly in the first several days of infection, and that a high percentage of symptomatic individuals were ill from some pathogen that was not SARS-CoV-2, producing false-positive results. Based on this information, it was unclear what an ill person with a negative lateral flow test result was supposed to do. Recommendations were to retest in forty-eight hours. In the meantime, millions of people were asking, "If I'm negative on my first test, do I need to quarantine from people I live with?"

More high-accuracy diagnostic tests did become available over time, but they needed to be conducted in a clinical lab. Often the combined time of going to a test center and waiting up to three to four days to get a result meant the patient faced many of the challenges a patient using the home lateral flow test encountered. One company, Labcorp, released a test on April 21, 2020, that enabled home, self-collected nasal swabs to be gathered and sent to their lab for analysis.[72] It could be days before the test subject learned the results.

On September 27, 2022, the FDA announced that 430 distinct tests had been issued EUAs for Covid,[73] and that moving forward, the

EUA focus would be on new technologies. In the meantime, a number of the companies that received millions in government support folded.[74] The commercial licensure approvals of these low-accuracy tests should be withheld in the future, and regulators must resist societal pressure to release such tests that produce high levels of false-negative results and do more harm than good—yielding countless millions of false-negative results actually resulted in unknowing disease transmission.

Covid showed that a diagnostic test's sensitivity as characterized by its "limit of detection"—the lowest concentration of measurable virus in a sample—is a key indicator of its performance, and better at detecting early infection than those noted in the manufacturer-conducted clinical trials. These trials included many individuals with significant later-stage infection, which is much easier to detect because of higher levels of virus but beyond the period when drug therapy can be most effective and after transmission of the virus can occur. Limit of detection should be a significant consideration in a diagnostic test's licensure review, and its measurement scale should be standardized to facilitate ready comparison between tests. Fundamentally, new diagnostic technology approaches are required. During times of epidemic/pandemic need, if the technology is not ready to scale, it's too late, and no amount of emergency funding or rapid procurement will alter this fact.

Before the Big One, we need a revolution in laboratory science and practice. We need testing that can be done at "point of care," or "point of need," including in the home. This type of test must be highly sensitive, at least equal to lab-based PCR testing. It should be easy for the user to place a saliva or other sample on a small device and within a few minutes get a result that is highly reliable in testing for various infectious agents at once. For example, a respiratory infection test should be able to test for ten to twenty of the major infectious agents causing respiratory illness. And this result should be able to be shared with one's healthcare provider. Imagine a physician now able to immediately prescribe an effective antibiotic or antiviral based on the test results.

I currently serve as a consultant to a company called GRIP Molecular that is working on this very type of testing. I enthusiastically agreed to assist because they are pursuing diagnostic approaches that check all the boxes necessary to revolutionize medical diagnostic testing as we know it. They intend to replace chemical assays, which are the norm today, with novel sample preparation and assessment methods yielding the accuracy of laboratory-based PCR tests in minutes. Their tests are intended to be performed by anyone, anywhere, at any time.

This type of technology addresses the inherent shortcomings of existing chemistry-based solutions to deliver actionable medical diagnostic information, that is, the ability to rapidly and accurately identify the virus, bacteria, or parasite causing a patient's illness, as well as potential antimicrobial resistance, so that the appropriate antimicrobial agent can be prescribed when it is most effective. This can all be done at the point of need, such as home, school, border crossing, eldercare facility, retail pharmacy, clinic, or hospital. The results can be transmitted to others directly via cell phone. The test result and other relevant information also can be provided, in a de-identified manner, to local health departments to enhance disease surveillance and effective public health measures.

This is the kind of innovation and creative thinking we desperately need as we contemplate the Big One. But it is not the only type of creative thinking we need. As we'll see in the following chapters, effective and accurate communication and rational, evidence-based political leadership will be equally important tools to get us through the Big One.

TAKEAWAYS

1. To be effective against the Big One and all future pandemics, we have to be prepared ahead of time. This includes properly resourced development of game-changing or universal vaccines,

antivirals, anti-inflammatory drugs, and rapid and accurate tests. This should be attainable with sufficient effort, funding, and realistic expectations for timing.

2. We don't wait for a war to develop and then build tanks, ships, planes, and other weapon systems. As the military is driven and funded by the inevitability of future conflict and need for protection, the war against infectious diseases should be similarly regarded, keeping in mind that the threat of serious harm and death from a pandemic is far greater than that of any conflict, short of a thermonuclear war.

3. Diagnostics must be seen as the prerequisite to effective and appropriate therapeutics, which in turn are a prerequisite for avoiding unnecessary illness, hospitalization, and death. We *must* know who is testing positive within the first few days of infection so that appropriate medical care can be given.

4. We can't count on the private pharmaceutical industry to bail us out in the next pandemic; the business model simply doesn't work. Governments, preferably of all the high-income countries, must be involved in funding research and purchase guarantees worldwide.

CHAPTER FIVE

EFFECTIVE COMMUNICATION

This is the West, sir. When the legend becomes fact, print the legend.
—From the 1962 film *The Man Who Shot Liberty Valance*

[Pandemic Month 6]

There were only four of them in the West Wing's windowless Roosevelt Room: the president and chief of staff Gilbert Stern in the middle of one side of the massive table; HHS secretary Anne Sulbarry and press secretary Deann Morgan on the other.

The president was dealing with the pandemic on multiple levels. He'd lost friends who had died; others remained seriously ill. It was restricting his actions and the lives of his children. He hadn't been sleeping well. It took more of his time and energy than he could afford to give any single issue. And he had seen his polling numbers steadily decline.

"We're being hammered by the media," he declared. "They say we're finally rolling out the new mRNA vaccine just as the pandemic is winding down. And it's true, things are getting better and case numbers are dropping. Right, Anne?"

"We're pretty sure it's just a temporary drop from last month's peak," Sulbarry replied. "Most of our experts think it's going to jump again, even higher than it's been so far."

"Experts! How is it our *experts keep saying things like 'We can't really be sure' and changing our strategies and mandates, and the experts on the other side are eating us for lunch? This week, it seems like that California doctor, Peter Minekal, has been on all the network and cable shows saying we're overstating the seriousness of SARS-3, no one should take the risk of getting the vaccine, because the pandemic is almost over, and any lingering, long-term effects people think they've got can all be explained as other diseases."*

"He and people like him said the same things during Covid-19, and they were wrong," Sulbarry pointed out.

"Well, why are all the media outlets putting him on?" the president demanded, clearly frustrated.

"Because he's a doctor, and he sounds authoritative," Morgan replied.

"He's a vascular surgeon, not an infectious disease specialist," Sulbarry clarified.

"Still, the media seem to like his delivery, and he's popular on social media because the anti-science/anti-vax movement in this country remains strong."

"We can't even get the vaccine out to everyone who wants it," the president said. *"And we still don't know how long the immunity will last, do we?"*

"No," Sulbarry conceded. *"But we're encouraged by the animal studies and clinical trials."*

"So, you put all this together, and even the people who aren't anti-science or anti-vax are confused. I can't say I blame them."

"Why can't we sound as authoritative as people like Minekal?" Stern asked. *"We're getting destroyed on social media."*

"Because there's still so much we don't know, and we don't want to give out information that isn't evidence-based and factual," Sulbarry said wearily. She didn't know how many times she'd had variations of this conversation since the virus first emerged — with reporters, government officials, even some of the HHS comms people. *"Much as we'd like to be, we're not driving the bus,"* she added. *"The virus is. All we can do is respond. And no matter what some people say, it's evolving faster than we can learn about it."*

"I know it's not what you want to hear," Morgan said frankly, *"but we believe that in the final analysis, calm honesty and humility about what we know and don't know is the best approach."*

"*How are we supposed to do that when people in the states where governors have shut down schools are complaining that their kids are falling behind, while in the states where they've stayed open, people are complaining that their kids are getting infected?*" the president countered. "*You saw how many turned out in Tallahassee at the memorial for that boy who died after catching the virus at school.*"

"*We're not sure that's where he became infected,*" Sulbarry noted.

"*Doesn't matter!*" the president said. "*It's still horrible... and bad optics.*"

"*I have to agree,*" said Stern.

The president continued: "*In Michigan, they stormed the statehouse because the governor instituted vaccine mandates for state employees. A home plate umpire died a week after calling a game in the World Series, and one of the catchers who was exposed to him was in intensive care for twelve days. Still, neither college nor professional sports decided to close down to fans this time around...*"

"*But in the hardest-hit cities, fans are mostly staying away from basketball games and hockey matches,*" Sulbarry pointed out. "*The arenas in those places are ghost towns. And even though they play mostly in open stadiums, we're seeing a lot of transmission we can link to football games. Just a month ago, nearly a third of the Sunday games were canceled or postponed because so many players and coaches were sick.*"

"*That's my point! People want to return to normal life, and we look like we can't control anything. A million, four hundred thousand deaths so far— what's that, like, ten times what was reported at the same point in Covid?*"

"*Seven times, but you're right, this is much worse,*" Sulbarry confirmed.

"*With all the supply-chain slowdowns and screw-ups, the news is showing stores with empty shelves. People are running out of basic necessities like toilet paper and aspirin! Drugstores and food stores are being looted. How do we get ahold of all this so we don't look incompetent or helpless?*"

"*I have to repeat, sir, a lot of this is simply out of our control,*" Sulbarry said. "*We're fighting a war with limited weapons. We're working on new testing protocols to get a better handle on localized intensity of the virus, but what we have so far under Emergency Use Authorization, we just don't know yet how accurate or reliable they are.*"

"*And why is that, six months into this thing?*"

"We didn't put the resources that we should have into developing new and better vaccines after Covid-19 so we'd be ready for this one. I wish I could give you a better answer, but that's the reality."

"So, what do you suggest?" the president asked, his tone betraying a hint of desperation.

Sulbarry and Morgan exchanged glances.

"What?" Stern said.

"Under the circumstances," Morgan began cautiously, "we think you should take a page from President Roosevelt during the Great Depression."

"I'm listening," the president said intently.

"Back then, despite all the new government programs, things didn't get much better economically for a lot of the country for several years. And just like now, there were only limited things the president could do. One thing he could do, though, was show his empathy for the average person and stress that he understood and that the government was doing absolutely everything it could."

"The Fireside Chats," the president stated.

"Exactly. Roosevelt got on the radio, with honesty and humility, and compassion, seeming to establish a direct relationship with every listener. If we schedule you for a series of 'chats' on the major networks and cable channels where you explain and level with the people—possibly with Anne and Tamara Goldfield to explain the technical details—we think that would go a long way to establish trust and authority."

"I agree with Deann," Sulbarry said. "Good, honest communication should always be the basis of public health."

The newsroom was on the third floor of the network's Manhattan headquarters, just above its two-story studio complex. In a glass-walled conference room across the aisle from the first row of reporters' desks, executive producer Herb Allen was going over the topic list for The Evening Roundup with news director Damon McNeil and Lisa Bryce-Colvin, who'd worked her way up from a summer internship to be the program's chief booker.

"What have we got for the pandemic package?" Allen asked.

"We've got Peter Minekal. In the pre-interview, he said he would talk about how the new vaccine may not be safe for everyone, and how there's *really no need to take it, because the pandemic is ending,"* Bryce-Colvin reported.

"Why are we having him on again?" McNeil protested. *"He's been wrong repeatedly."*

"He's a vascular surgeon. Our viewers pay attention to him. And he's very self-assured," Bryce-Colvin said. *"How do we know he's not right? He said case numbers would go down, and they have. And he's a bestselling author."*

"On nutrition and supplements for heart health," McNeil countered, *"which have nothing to do with infectious disease epidemiology. Why don't we get Dr. Robert Andrews from the University of Michigan?"*

"Bad News Bob? He scares the hell out of everyone," Allen said.

"But he's been right," McNeil said. *"He predicted the number of deaths we'd have from Covid-19. Everyone said he was a doomsayer, and he turned out to be right on the money."*

"He can be such a downer, though," Allen said. *"Do we want to depress our entire viewing public? Look, we're not doctors or scientists. Doesn't Minekal have as good a chance of being right as Andrews?"*

"Not if his track record is any indication," McNeil responded.

"Look," said Allen. *"Let's just go with Minekal and see how it goes. Lisa, tell him he's on during the eight-o'clock hour."*

———————

At 8:20 that evening, Hannah Cipora, one of the Roundup *hosts, introduced Dr. Peter Minekal, sharing the split television screen with him.*

"Dr. Minekal," she began, *"despite your well-known pronouncements, wouldn't you say that a mortality rate of ten and a half percent of those afflicted by SARS-3 is too great a devastation to ignore?"*

"If that were the case, I might agree with you, Hannah," Minekal responded. *"But I don't see any hard evidence that the mortality rate is anywhere near that high."*

"Are you saying these people didn't die?"

"No, of course not. What I'm saying is most of these deaths are the result of some other cause or underlying morbidity. For example, if someone dies in a car crash, and he or she is found to have coronavirus infection, you could say they died of SARS-3."

"Are you really willing to use that as an example, Doctor?"

"Okay, perhaps that is a bit extreme, but you get my point: I believe that many deaths have been attributed to SARS-3 merely because the virus was present in the bodies of those who died."

"Do you have any hard evidence of that?" Cipora challenged.

"I think it should be obvious," he replied.

"Well, it's apparently not obvious to a lot of experts and public health professionals. Dr. Robert Andrews, with whom I know you have publicly disagreed on numerous occasions, has predicted that if this pandemic lasts in some form for three years or more, as Covid-19 did, we could easily be looking at seven to eight million deaths in the United States, and hundreds of millions worldwide."

"Clearly, I don't accept that, Hannah, particularly when all indications are that the outbreak is on the wane and case numbers are declining almost everywhere."

"I notice you said, 'outbreak' rather than 'pandemic.' Are you sticking to your assertion that for the last almost six months we have not been in the midst of a pandemic?"

"We can play with words all you want, but that's not going to get us any closer to the truth."

"So, to be clear, are you denying the existence of the pandemic?"

"Let's move on," Minekal said. "I think your viewers and my readers and followers have more important things to think about than nomenclature."

"Okay, Dr. Minekal, to get back to Dr. Andrews's predictions, you don't accept that this pandemic could kill more than seven million people in the United States alone?"

"I do not."

"Even though, while others were saying that the worst of SARS-CoV-2 was behind us, he said, in an interview with journalist Paul Babcock in April 2020, that the darkest days were still ahead of us? At that time, he predicted we'd have eight hundred thousand deaths in the US in eighteen months, and he turned out to be right."

"Well, I don't accept that Andrews was right. Did we do autopsies on all those people? We really don't know what they died of, but I don't think the vast majority died of Covid-19. And most of the ones who did had underlying conditions that made them particularly vulnerable. But that's all history. Today, *the infection rate is going down, case numbers are going down, and soon we won't have to worry about the virus. You can trust me on that!"*

Sitting in the den of her home in suburban Minneapolis after a typically long day in the ER, Dr. Erin Thomas pointed the remote at the television and, with more force than needed, decisively clicked it off. She couldn't stand to hear people like Peter Minekal spouting opinions that had already been proven wrong time and again. Why do the networks keep putting him on? *she wondered.*

She sipped the glass of wine she'd been nursing since she got home from her shift. She knew she really should watch how much she drank, but after six months of round-the-clock crisis-management mode in the ER, a drink had become almost a nightly necessity. Every day, dealing with bad outcomes that could have or should have been prevented: heart attack, stroke, and childbirth complications in people who didn't come to the hospital for fear of catching SARS-3 until it was too late; accident victims who overwhelmed the already strained resources of the emergency department. She thought of the forty-five-year-old man she saw earlier that day—the father of several young children. She'd suspected he had blocked arteries, but before he could be treated, he had to be screened for the virus, isolation protocols had to be instituted, and they had to wait until the cardiac catheterization lab was thoroughly cleaned from the last patient before he could be brought in, costing precious time he didn't have. He coded before they ever made it to the lab, and they couldn't get him back. Losing any patient was devastating, but today she realized, as she faced his wife, she was growing numb to the grief, having just seen too much.

The workload was unrelenting. Forty-five of her staff had developed SARS-3, and twelve were still hospitalized or out of work with long SARS-3, which was even more debilitating than long Covid had been. Six hadn't made it. And her hospital was doing better than many, based on the rigid precautions she'd instituted in the ER, as well as sheer luck. She knew a lot of other people in the community, some of whom were close friends, who hadn't been so lucky. Thomas didn't know how much more heartache, how many more virtual memorial services, she could take.

But apart from her own friends and colleagues, perhaps the saddest cases—the ones that infuriated her at the same time they broke her heart—

were the terribly sick patients who'd been vehement anti-vaxxers right up until the moment they realized they were going to die. Then they would either beg for the vaccine or go into total denial. Either way, it was too late. She wished she could record those moments and show them to the protesters and crazies who kept harassing her on the way to the parking garage, not that she thought it would make much of a difference. Then she'd have to deal with family members, many of whom could hardly control their anger and anguish at their deceased loved one for not heeding their warnings. Some were irate, claiming their relatives couldn't have died from a "fake" disease. Everyone had their own reality, and they all wanted quick, unchanging, black-and-white answers. How could you convince them that that wasn't the way science worked?

OF ALL THE TOOLS in the public health arsenal, effective communication is among the most important. Viruses are complicated. But if the information released about a disease, and guidance on its prevention, is confusing, incomplete, or downright inaccurate, the consequences will be so much worse, as people won't know what to do or whom to believe. When we think about what the ordinary citizen can do to prepare for a pandemic, there generally isn't a whole lot other than advocating with local, state, and national officials for advanced planning and adequate resources. However, once a pandemic actually hits, the success of containment and mitigation efforts relies to a great extent on the actions and compliance of each of us in protecting ourselves, our friends and families, and our communities. That's why effective communication is so vital.

But Covid showed how quickly communication can break down or become counterproductive—thanks in part to a lack of shared understanding of the information being communicated.

Today, we have at least three buckets of public health information: facts, misunderstandings, and outright lies. Unfortunately, we have seen the amount of mis- and disinformation only increase. During Covid, this included statements attributing major safety risks to the use of vaccines and promoting the effectiveness of drugs like ivermectin to treat critically ill Covid patients.

The disinformation could have tragic real-world effects. In early January 2024, Politico reported, "Nearly 17,000 people may have died after taking hydroxychloroquine during the first wave of Covid-19, according to a study by French researchers. The anti-malaria drug was prescribed to some patients hospitalized with Covid-19 during the first wave of the pandemic, 'despite the absence of evidence documenting its clinical benefits,' the researchers point out in their paper, published in *Biomedicine & Pharmacotherapy*."[1]

In summarizing the finding, the Kaiser Family Foundation's (now KFF's) *Health News* commented, "At that time in the covid pandemic, then-President Donald Trump said of the unproven treatment: 'What do you have to lose? Take it.'"[2]

Aside from downright dangerous misinformation, evidence based on rigorous investigation, study, or instruction is often available, but it is relative to a given context and time frame. For example, the guidance for protecting oneself issued to people in a large city in the middle of a wave of transmission is different from what we'd say to those in an area where fewer cases have been reported. As another example, the Covid pandemic evolved from 2020 to 2023 as the mutated variants impacted the transmissibility and the spectrum of serious illness the virus caused. Information collected—using scientifically backed study methods—on what SARS-CoV-2 might inflict on humans changed substantially over time. These changes led some to believe that public health officials' statements about Covid were not truthful. That is one problem.

On top of that, pandemic communication isn't a one-lane road, or even a two-way street. It's a multilane superhighway interchange, with traffic zooming past in every possible direction.

There are national, state, and local government leaders, plus a multitude of international, federal, and state agencies, including WHO, HHS, the CDC, the FDA, the NIH institutes and centers, the Surgeon General's Office, and the National Institute of Occupational Safety and Health, as well as the 50 state and more than 3,300 local health departments.[3] Sometimes, one arm of a given agency contradicts what another arm is putting out. People get their information from social media, television and radio, newspapers, magazines, podcasts, blogs, newsletters, and more. There are all the talking-head professionals, each with his or her own perceived level of understanding, agenda, and curated set of facts, including many who don't know what they're talking about, others who are consistently wrong, and some who are simply off the wall. In any crisis situation, the media at large becomes the equivalent of air traffic control, determining who and what gets to the public, and how frequently. Add to that the quick appointment of reporters to the pandemic beat who had little or no grounding in science, and all the social media posters who have even less experience and are intentionally or unintentionally spreading false information and groundless rumors.

So, when we talk about lessons learned from the communications failures of the Covid pandemic, who of the countless communicators are we talking about? That's not an easy question to address, and it would be monumental hubris for us to claim to have all the answers. How can I both criticize the national health institutions for failing to speak with one collaborative voice and at the same time say that guidance depends on context and time frame, which requires continual flexibility in messaging? I can only say that this apparent dichotomy goes to the essence of the challenge we face.

When people mention the "public health establishment" to me, I know I'm included in that designation, yet I have disagreed publicly with its leaders on multiple occasions. We have to acknowledge that we don't all speak with the same voice. Still, effective communication is so important in a pandemic that we must try to be better and more consistent communicators. And, just as important, we must be better listeners.

At its core, we are talking about effective risk communication. Here we might ask, What exactly is risk communication in a pandemic? WHO defines it as "the real-time exchange of information, advice and opinions between experts or officials and people who face a hazard or threat to their survival, health, or economic or social wellbeing. The purpose of risk communication is to enable people at risk to make informed decisions to mitigate the effects of a threat (hazard)—such as a disease outbreak—and take protective and preventive measures. Risk communication is proven to be a critical tool in emergency preparedness and response."[4]

The United States got off to a bad start with Covid-19. President Donald Trump publicly minimized the threat of the virus, even though it was later learned through author-journalist Bob Woodward's reporting that he was aware of its seriousness.[5] Official voices, both within the United States and globally, offered differing and often confusing guidance. We addressed in Chapter Two how frustrating and dangerous it was that throughout the pandemic, authoritative organizations such as WHO and the CDC would not acknowledge that the virus was primarily transmitted through aerosols, despite evidence that was clear and compelling to many of us in public health. There was no coordinated messaging. Political figures who knew little or nothing about medicine, virology, or epidemiology offered their own opinions on treatments that were, in fact, useless or actually harmful. For example, some asserted that there were simple fixes, like vitamins or other supplements, but that Big Pharma was suppressing the information because there was no profit in it for them.

And then there was the "mask" debacle. Unless we get agreement ahead of time that we will follow the Precautionary Principle with any unknown (particularly rapidly spreading) disease, we will likely face a similar predicament when the Big One hits.

When Covid struck, the CDC advocated testing, but it proclaimed that only the tests it developed were acceptable and discouraged private labs from coming up with their own. Then, when the CDC-sanctioned tests turned out to be defective,[6] we lost weeks of critical data, and with it, the ability to figure out who most needed attention and treatment. These and other examples of poor communication ultimately fed distrust in the government and the scientific establishment and contributed to the cascade of mistrust, anti-vaxxers, and anti-science conspiracy theorists who turned every public health decision into a political proxy and litmus test. While politics often influences response to a virus, the virus is apolitical.

We have stressed that a critical aspect of pandemic communication is that it stay current, with messaging evolving as the conditions on the ground change. This was a critical failure of communications during Covid: Messaging didn't change much; it didn't keep up with the times and the way the virus itself was evolving. Part of that was due to the fact that our frame of reference for a pandemic was with influenza: in 1918, before we'd even discovered it was a virus; and in 1957, 1968, and 2009, when there was limited understanding of how viruses change over the course of a pandemic. With this as background, we were unable to understand that a pandemic virus could mutate or evolve significantly over the multiyear course of the pandemic. Scientists and government officials simply were not prepared for the evolution of the SARS-CoV-2 coronavirus variants. Equally significant, however, was a general reluctance to seriously consider the evidence and constantly reevaluate the approach and messaging as reality dictated.

As Mark's late friend, the acclaimed espionage novelist and former CIA clandestine operative Charles McCarry, observed, "Genius is the ability to perceive the obvious." The inability of those in authority to perceive the obvious during the Covid pandemic led to a lack of public trust in institutions that we desperately needed to be effective.

More than five years of doing the *Osterholm Update* podcast during that pandemic confirmed for me what the essence of effective public health communication needs to be. First, it has to be scientifically accurate. The questions then are, by whose standards, and what evidence is needed? At CIDRAP, we regularly evaluate a multitude of research studies. We look at how each is designed and carried out, and at factors such as how many patients were involved, to assess statistical significance. Are the findings supported by the data? Have the results have been replicated in other studies? These are among the generally accepted parameters applied to evaluation of proposals for research funding or publication in peer-reviewed scientific journals.

One challenge is that even when we go through this evaluation process, it doesn't mean we will get all the answers. But we must apply such standards and then present the most up-to-date scientific information, based on the evidence and data that we have, while acknowledging that it is likely to change or evolve as we learn more. This is the juncture where expertise and genuine humility meet—what we've characterized as "calling balls and strikes."

Here's the next challenge: In the feeding frenzy for information that hits during a pandemic, the media (and others) tend to jump to publish and post content as soon as it is available, even before it has been peer-reviewed. We saw this with the release of "preprint" abstracts on research not yet peer-reviewed. Preprints are essentially summaries of research articles not yet cleared for publication. Anyone can submit a preprint to one of the services that then posts it online. While it seems a good idea to communicate scientific or medical findings as quickly as possible in a pandemic, everything must be critically evaluated—as we do at CIDRAP, and as the peer-review process typically works. A subject matter expert will understand how much weight to give a preprint. But the media, with little to no expertise in the specific issue or topic, in many cases can do no better than report, and they seldom follow up to see whether peer review bore out the conclusions they reported. Some Covid-related findings publicized on the basis of preprint information later proved to be flawed or incomplete, the conclusions couldn't be verified by other data or studies, and the preprint never advanced to publication of a full article in a scientific journal. But the genie was already out of the bottle, and the public knew only what they'd originally read. This is likely to continue to be a difficult problem going forward.

We have to make clear that science is an iterative process whereby we implement something based on what we know so far, then we study it and learn about it. And what we learn then gets incorporated back into the implementation to improve on whatever it is we're trying to do. And then we study that, and once again learn from the subsequent implementation, and we keep getting better with each one. I wish we could have perfect vision to know the exact science in any one given moment with clarity and a comprehensive view, but we don't, and we won't. It is always going to be an iterative process of implementation, studying, learning, and re-implementing. This is why I call it "evolving science," and it is part of our job to help the public understand.

Just as important, we have to relate to the public as a community, to let everyone know that we're all in this together and acknowledge that it is going to be hard but that we're going to get through it. The model for this is President Franklin Roosevelt's Fireside Chats, the approach White House press secretary Deann Morgan recommends to the president in our scenario. It is critical for leaders, whether they're in government or health and science, to show that they honestly care; that they understand that every data point or line on a graph represents a human being. We may not be able to relieve all suffering, but we can always care, and show it.

In September 2001, after the anthrax attacks in Washington, DC, and Florida, which followed the 9/11 horror by only a week, Mark coproduced and wrote, and I participated in, a program for PBS entitled *Bioterror: Coping with a New Reality.* There was certainly reason to be afraid then, which we acknowledged. But in the closing of the program, host Scott Simon reminded viewers that during the Battle of Britain in the fall of 1940, Londoners endured months of vicious bombing from Hitler's Luftwaffe, often sleeping on subway platforms, yet still retained their basic humanity and, to the extent possible, went about their daily lives. When people face a common threat or challenge together with the assurance from leaders that they will collectively get through it regardless of the pain and suffering, they often surprise themselves by their ability to rise to the occasion.

When I was the state epidemiologist in Minnesota, like "Bad News Bob" in our scenario, the local media sometimes referred to me as "Bad News Mike," saying a call from me to a government official was rarely good news. While it was often true that I was the bearer of ill tidings, I've always felt that leveling with people in a collaborative relationship to deal with the challenge was not only the most effective strategy; ultimately it would be the most reassuring. When I get pushback for speaking the truth as I see it, I always say that I'm not trying to scare anyone *out of* their wits, I'm trying to scare them *into* their wits. You can't make decisions, plan, or act without accurate information, and I've never understood how you can gain anyone's trust by not telling the truth or by withholding critical information.

This was a real challenge for me during Covid. At the end of January 2021, we were seeing the overwhelming effects of emerging variants in other parts of the world. In England, for example, the B.1.1.7 variant (later referred to as "Alpha," when WHO started using letters from the Greek alphabet to make variant names more accessible) sent twice as many people to the hospital as we'd ever had hospitalized in the United States at our highest numbers—and it was just starting to hit the United States. At the same time, we still had a vaccine shortage.

From a planning standpoint, I felt it was critical that people realistically understood the threat we were facing. I appeared on *Meet the Press* with Chuck Todd on January 31 and said, among other things, that it was time to call an audible (using the American football expression) and focus on getting initial doses into as many people aged sixty-five and older as possible, to reduce the "serious illness and deaths that are going to occur over the weeks ahead." I warned, "I'm telling you, the darkest of the darkest days are yet ahead."[7] I took a lot of heat for that statement—from the media, politicians, and even colleagues. But I could see what was coming down the pike.

In fact, the darkest days of the pandemic *did* occur after my January 31, 2021, statement. From the beginning of the pandemic until January 31, approximately 490,000 Americans died from Covid. From February 1, 2021, through the end of December 2023, about 680,000 additional deaths occurred.[8] The number of Covid patients in US hospitals peaked in late January 2022, at almost 160,000.[9] I am haunted by the thought that some might have been saved if we had faced—and prepared the public for—the truth in January 2021, instead of having so many in authority falsely reassure people that things were getting better.

The goal of sugarcoating may be to keep people calm. If so, though, it typically backfires. Over-reassuring messaging makes frightened people feel abandoned, alone with their fear. They smell a rat, even if they don't know exactly what frightening facts you're leaving out or papering over, which makes them even more frightened. They also trust over-reassuring officials less and become less willing to follow their lead.

By contrast, candor about present or impending bad news can be paradoxically calming. We feel not only well-informed but well led, and more confident that our leaders aren't treating us like children. It's bracing. It's the other shoe dropping. And often the bad news isn't as bad as our worst fears.

Even leveling with the public that the problem has not gone away and that huge challenges remain can be put in a positive or reassuring context with proper leadership. I am often reminded of and frequently quote Prime Minister Winston Churchill's message just after the Allies defeated the German and Italian forces in the Battle of El Alamein. That was in 1942, and Churchill knew that a long, hard, and bloody war still lay ahead. Yet he was able to rouse and reassure his people when he declared, "Now this is not the end. It is not even the beginning of the end. But it is, perhaps, the end of the beginning."[10]

THE THREAT MATRIX

Another challenge we face in public health is that people generally don't always have a rational sense of what the real threats are. In other words, there is often a mental/emotional disconnect between what is most likely to kill us, what can hurt us, and what mostly just makes us feel uncomfortable. We refer to this concept as the threat matrix.

An accurate threat matrix would be a graph that shows us what we should reasonably be worrying about. It might have a vertical axis representing severity of a particular risk and a horizontal axis representing probability of occurrence. In a pandemic, it is critical for the public to understand the threat matrix and be able to apply it to their own lives and situations so they can make informed decisions. Communicating a realistic threat matrix is a responsibility leaders should embrace.

The inherent problem is that most people's personal threat matrices are not accurate in that they don't or can't objectively evaluate degree of risk. Most of us know, for example, that commercial airplane travel is far safer mile for mile than automobile travel, yet many of us fear flying but get in our cars every day without a worry. Part of this has to do with our inherent fear of lack of control in a plane versus a car, and part has to do with the familiarity of one mode of transportation versus the other.

Here's an example of this disconnect in public health: During the 2014 Ebola outbreak in three West African nations, Mark and I attended a breakfast on Capitol Hill to advise senators, representatives, and their staffs on the relative threats the virus posed to us here in the United States. Many of those in attendance, particularly on the conservative side, wanted to shut down international aviation and commerce from Africa and impose some severe restrictions here in the United States. I tried to communicate that the very nature of the disease—when it became infectious, how it was transmitted—and the resources we had to deal with it prevented it from becoming epidemic in the United States, despite the tragic experience in West Africa. Maybe their reaction had something to do with the grisly (and largely overblown) descriptions of bleeding eyeballs and organs turning to mush. When it came to SARS-CoV-2, which really *did* present a serious public health threat here, much of this same cohort didn't want to do anything to restrict commerce and protested that any health-related mandates were an infringement on personal liberty. It's the same phenomenon as focusing on the risk of getting vaccinated without understanding that the risk of not doing so is far, far greater.

When Covid-19 vaccines became available, we heard from women of childbearing age that they were worried by stories they'd heard that the vaccine could diminish fertility,[11] then and in the future. This was certainly a reasonable concern, so when compelling data were available to dispel that fear, it was incumbent on public health authorities to convey that information, as well as the risks associated with giving birth while suffering from SARS-CoV-2 if a woman had not been vaccinated.

With a vaccine, there are some slight numerical risks, which can be accentuated in certain circumstances if an individual has an underlying medical condition or comorbidity. This will likely be true during the Big One, assuming we have developed effective vaccines. But it will almost certainly also be true that the realistic threat matrix will overwhelmingly favor the slight risk of being vaccinated with the severe risk of contracting the virus if unprotected. It is vital that we explain and communicate the threat matrix concept as a normal component of public health communications—starting now, before the next pandemic hits—so that when the crisis occurs, the concept will already be familiar, and individuals can competently evaluate their own threat matrices in making informed decisions.

I'm not relitigating the idea of mandates here, nor am I suggesting that all vaccines are either perfect or equal. I'm only suggesting that before the Big One, we have to come up with a better way of communicating actual risk versus its emotional correlative in a way that makes sense to most people and doesn't threaten their political views or suggest an ongoing danger to their freedom. I don't know if that's possible in our current climate, but it's too important not to try.

LEARNING FROM EXPERIENCE

We have already shown how even official sources such as WHO and the CDC let erroneous or outdated information remain on their websites. Unless there is greater oversight and more responsible reporting within those organizations, we won't have substantially better crisis communication during the Big One than we had during Covid-19. But what would we like to see happen?

Dr. Peter M. Sandman is a leading risk communications specialist, retired professor at Rutgers University, and founder of the institution's Environmental Communication Research Program. Together with his wife and colleague, Jody Lanard, Peter has made valuable contributions to the study and practice of public health communication. Peter and Jody have been trusted advisers of mine for many years. Peter has identified eight risk communication mistakes that government and public health leaders make that contribute to the ineffectiveness of pandemic messaging and sap trust in the information and advice we give out. "Turning them around," he comments, "can build trust and save lives."[12] Many of the mistakes he's pinpointed touch on concepts we examine throughout the book.

Overconfidence and Failure to Proclaim Uncertainty

Again, this is where humility comes in. People understandably want definitive answers and messaging that doesn't change. But since that's not how science works, Peter recommends not just *acknowledging* uncertainty, but *proclaiming* it. He states, "The public can tolerate official uncertainty, if it's confidently and matter-of-factly stated."[13] As examples of overconfidence without acknowledging uncertainty, he cites the Trump administration and public health officials' assertions that the SARS-CoV-2 virus was not spreading significantly in the United States; that masks were not needed; that washing hands and cleaning surfaces were the most important means of preventing spread; and failing to realize that transmission was primarily airborne.

Reflecting on how scientific authorities often missed the boat, former NIH director Dr. Francis Collins, speaking to a group of journalists in September 2022, humbly admitted, "The big thing that I know I didn't do, and I don't think a lot of the communicators did, was to say this is an evolving crisis, this is going to change every time we made a recommendation, whether it was about social distancing or mask wearing or vaccines."[14]

I admire the way Francis owned up to his own overconfidence, and I think it's something we can all learn from. People seem more comfortable saying something with apparent authority than admitting they don't know. I'm always struck by the way so-called financial experts who haven't a clue what the stock markets will do tomorrow (if they did, they and all their clients would quickly become multimillionaires) will go on television and explain exactly why the markets closed the way they did today. In the same way, too many government and public health officials will try to explain trends even though, as Peter says,

> They don't have a clue why something happened; that the virus does what the virus does; that we're not steering this ship, we're passengers....
>
> When you overconfidently attribute events we don't really understand, like rises and falls in case numbers, it undermines public confidence in the things you truly can confidently attribute. The fact that vaccinated people are much less likely than unvaccinated people to be hospitalized with Covid is genuinely attributable to vaccination, for example.[15]

Falsely positive messaging, on the other hand, is likelier to backfire than to inspire. President Donald Trump justified some of his overly optimistic briefing claims on the grounds that cheerleading is part of a president's job.[16] That's true when the country has a winning Olympics team. But in one of the deepest crises since World War II, what's called for is a Churchillian combination of transparency and determination, not unrealistic cheerleading.

Failure to Do Anticipatory Guidance

Telling people what to expect is a linchpin of crisis communications, and it's what Peter means by anticipatory guidance. Realistic worst-case scenarios, presented by reasonable, rational, and trusted authorities, can help neutralize even worse possibilities in the minds of a terrified public. On the other hand, we lose vital credibility if we merely try to calm nerves. Peter notes, "One of my favorite good examples is from CDC head Jeff Koplan in the early days of the 2001 anthrax attacks: 'We will learn things in the coming weeks that we will then wish we had known when we started.' That's an elegant way of saying, 'We expect to make mistakes, find out about them, tell you about them, and correct them.'"[17]

Giving anticipatory guidance also prevents people from being taken by surprise, which, in itself, can also lead to mistrust. For example, an official responsible for a given geographic area could say that if positivity rates for the virus go up to a certain level, schools may have to be closed for a limited period and N95s will be required for indoor public spaces.

Dr. Sandra Quinn is chair of family science and senior associate director of the University of Maryland School of Public Health Center for Health Equity. She has done considerable research on the best ways to communicate vital information during a pandemic. She says, "In an ideal world, part of what we spend a lot of time doing is preparing people for what is coming.... Early on [with Covid-19], smart leadership would have convened people who do this kind of communication to talk about how we communicate: 'This is coming.... There will be the issue of uncertainty....' We would have started talking about what we know: 'There may be a number of ways we're going to address this.' That could be something about masks or high-quality respirators as a means of protection, the potential for social distancing or reducing our contact with others.... We knew our evidence wasn't that strong, but we make an assumption that you have a contagious virus, and if you put a lot of people together, we know that's a problem.

"That [also] includes starting with legislators early on: 'This may require a number of measures that may be uncomfortable. We don't know which we will need, in what time frame, but we'll keep in touch and be updating you.' That's hard to do in an alternate-facts, social media environment [where people] express things as pronouncements, not as a conversation." She believes if there had been a different approach from the beginning, "of talking and helping prepare people about what was coming, and then updating them about why we now have to make this decision, and that it will affect different people differently... [the negativity] would have still been there, but it would have been offset a lot."[18]

It also helps to work within communication channels already established, respected, and — most importantly — trusted in a community. "Community" may be defined by geography, ethnicity, religion, sexual orientation, immigration status, or any other uniting factor. We've seen from the early days of HIV/AIDS, for example, how identifying trusted sources for information within a community can be critical in engaging the population in disease prevention.[19] Religious leaders can be trusted messengers, and it is unfortunate that in the Covid pandemic so many were alienated, whether by disinformation about the vaccine or by restriction of large gatherings in places of worship. Before the Big One, we must establish a communication network that flows from the highest levels in global and national public health, scientific research, and medicine, all the way to the community level, identifying and coordinating with trusted messengers on a regular basis. This communication should be two-way, as these messengers are best able to understand the concerns of those in their community, enabling them to serve as spokespeople both up and down the chain and to be proactive.

It is vital that information be made available to everyone. Among the disparities highlighted by Covid was the fact that internet access is not universal, even in the United States and other developed countries. We can't yet abandon older means of communication. For example, in some places, a billboard is the best way to reach people. It seems obvious that messages need to be translated into the languages spoken in different areas, but we need to consider this in all communication. In following up with people who test positive, the person(s) contacting them should not only speak their language fluently but understand any cultural or socioeconomic barriers to being able to institute appropriate measures.

Fake Consensus

With the myriad uncertainties that a pandemic will inevitably bring, there are bound to be people within the scientific and public health communities who disagree with the majority. If those voices are either silenced or not acknowledged, and any aspect of what they have stated turns out to be true, the credibility of authorities can be irreparably shaken. For this reason, Peter thinks that the traditional axiom to "speak with one voice" is a dangerously oversold concept. If there is true consensus within the public health community, fine, as long as it acknowledges the possibility of new evidence emerging that can change our understanding and directives.

I have experienced this phenomenon firsthand. During Covid-19, I found that my outspoken approach, which often put me at odds with the "authorities"—as when I asserted that the evidence showed the virus was spread primarily by aerosols rather than droplets; or when I could not agree with the overly optimistic assessments about when the pandemic would be over, for example—also meant that certain media programs didn't want to have me on, and certain publications didn't want my op-eds. This caused me no particular harm, but how much better for public awareness would it have been if all views were at least acknowledged, so people would understand which issues were still open to interpretation and subject to further evidence. Even self-styled authorities dispensing misleading information and advice, like Dr. Peter Minekal in our scenario, can be given a forum as long as they are countered by responsible, evidence-reliant authorities. You don't have to agree with someone to acknowledge their position and then give your own. To my way of thinking, this is a far better approach than simply ignoring them, because in today's media environment, their message is going to get out anyway.

An example of this during Covid was the Great Barrington Declaration of October 2020, sponsored by the libertarian American Institute for Economic Research.[20] A group of doctors, scientists, and academics, some from internationally prominent institutions, published an open letter calling for a more laissez-faire approach to SARS-CoV-2, protecting through some form of isolation or sheltering in place those who were at greatest risk for serious disease, like the elderly and immune-compromised, and letting the virus spread throughout the rest of the population. That, they asserted, would lead to faster herd immunity—within three to six months—and less economic and personal disruption. How we were supposed to protect those at risk who lived in homes with others who were going about their normal business, the Barrington authors didn't say.[21]

Predictably, many libertarians and those opposed to any mandates or closings embraced the declaration. Many of us in public health had the opposite reaction. WHO director-general Dr. Tedros Adhanom Ghebreyesus asserted that "herd immunity is achieved by protecting people from a virus, not by exposing them to it. Never in the history of public health has herd immunity been used as a strategy for responding to an outbreak, let alone a pandemic. It is scientifically and ethically problematic."[22]

On my podcast and in media appearances, I stressed that it was essential to keep transmission to a minimum as we awaited a vaccine, or hospitals and healthcare facilities would be overrun. The timing of the Barrington declaration was absolutely wrong—in October 2020, protection provided by vaccination was on the horizon. It made much more sense to wait for immunity via this mechanism than to risk the health and lives of so many needlessly. I also took issue with the declaration's claim that government orders were primarily responsible for the economic retreat. Rather, I said, it is because people didn't feel comfortable engaging in a multitude of public activities.

As it turned out, the basic premise of the Great Barrington Declaration was off base for two critical reasons. Contrary to what we all hoped (but many of us in public health doubted), the vaccine, as we've noted, did not prevent transmission.[23] And because of the evolution of variants and waning immunity from vaccination and/or having had the virus, herd immunity was an ever-receding goal. If you can catch the virus more than once, the concept of herd immunity is off the table. But the point here is that even a statement put out by seeming experts had to be questioned, and those of us who disagreed with it had to be vocal, even though the apparent lack of consensus within public health no doubt added to the public's confusion.

Prioritizing Health over Other Values

Public health has always been my number one professional concern. But that doesn't mean that medical considerations can, or should, be the only drivers of policy. Society is too complex for that. As Peter Sandman puts it, "Public health officials and experts understandably prioritize health over many other goals and values: liberty, property rights, education, economics, psychological wellbeing, convenience, quality of life, etc."

He adds, "In each instance public health officials are entitled to make their case that health should prevail. But they're not entitled to pretend that there is nothing of value on the other side of the debate."[24] Oftentimes there are trade-offs to be made, such as keeping schools open for the sake of ongoing educational goals even though some transmission will result. And as we suggested in Chapter Three about mandates, if they have to be imposed for the sake of widespread public health concerns, it's important to acknowledge that those imposing them understand that they are an infringement on liberty, or even bodily integrity in the case of vaccination, and that the balance has been carefully weighed.

In a recorded discussion hosted by the bipartisan grassroots organization Braver Angels in July 2023, Francis Collins noted, "If you're a public health person and you're trying to make a decision, you have this very narrow view of what the right decision is. And that is something that will save a life; it doesn't matter what else happens.... You attach infinite value to stopping the disease and saving a life. You attach zero value to whether this actually totally disrupts people's lives, ruins the economy, and has many kids kept out of school in a way that they never quite recover from....Collateral damage." He added, "This is a public health mindset. And I think a lot of us involved in trying to make those recommendations had that mindset, and that was really unfortunate. It's another mistake we made."[25] While his comments certainly don't apply to everyone in public health, they do highlight the dangers of strictly prioritizing medicine and science over social values and practicalities, and I commend him for his honesty and candor.

Prioritizing Health over Truth

In an effort to get the public to go for flu shots, for a long time the public health establishment stuck to the message that on the whole, the vaccine was 70 to 90 percent effective in preventing illness.[26] Yet in some flu seasons, the influenza vaccine was closer to 25 to 50 percent effective in preventing illness severe enough to seek medical care,[27] and as we've mentioned, immunity began to wane in two months or less.[28] While conceding this, I still advocate getting vaccinated, because it could at least cut down your chances of contracting flu and lessen the seriousness of the effects if you do get it.

In this case, as in so many others, it behooves us to follow the Churchillian adage: Communicating uncertainty is more credible than faking overconfidence.

Failure to Own Your Mistakes

This mistake pretty much speaks for itself, but it's vitally important. How much time, money, and human resources have been wasted on what amounted to hygiene theater? How many times have we cited directives on web pages and other sources that were never amended or corrected even after contradicting information or data were available? It's understandable that people in positions of power and authority don't like admitting they were wrong, and often they are trying to preserve their agency's credibility.

Yet it is completely counterproductive to refuse to admit and apologize for an error everyone knows you made. And the list during the Covid pandemic is long: Saying SARS-CoV-2 probably didn't transmit from human to human; saying asymptomatic people probably didn't transmit it; saying it was unlikely to spread much in the United States; saying we've got it under control; saying our hospitals had plenty of personal protective equipment. It's not difficult to remember or find out who said what. It's far better to say you were wrong and explain, if necessary, what went into your initial decision process and the reasons you have changed your mind. That is how you reestablish credibility.

I experienced this in trying to predict how the Covid pandemic would play out over the course of two to three years. It was unclear in the earliest days of 2020 how long a coronavirus pandemic might last; we had no previous experience with this virus causing a pandemic. Rather, our experience was based on four influenza pandemics of the previous 102 years, which lasted from sixteen months (2009) to three-plus years (1918.)[29] As noted earlier, we had no epidemiologic evidence that throughout the course of these pandemics the virus underwent major mutations, introducing strains that might make reinfections possible. It was generally assumed by public health experts that once infected, you either died or recovered and then were protected against another influenza episode, until your immunity waned.

In April 2020, Drs. Kris Moore from CIDRAP and Marc Lipsitch, director of the Center for Communicable Disease Dynamics at the Harvard T.H. Chan School of Public Health, 1918 influenza historian John Barry, and I coauthored a CIDRAP Viewpoint entitled "The Future of the Covid-19 Pandemic: Lessons Learned from Pandemic Influenza."[30] Viewpoints were a series of expert review papers CIDRAP developed specifically to discuss critical aspects of the Covid pandemic. In the April 2020 document, we laid out possible scenarios for how the pandemic would evolve and ultimately end. We assumed that at some point we would reach herd immunity, a status where 60 to 70 percent of the population would be immune from previous infection or vaccination. However, we stated,

> This may be complicated by the fact that we don't yet know the duration of immunity to natural SARS-CoV-2 infection (it could be as short as a few months or as long as several years). Based on seasonal coronaviruses, we can anticipate that even if immunity

declines after exposure, there may still be some protection against disease severity and reduced contagiousness, but this remains to be assessed for SARS-CoV-2. The course of the pandemic also could be influenced by a vaccine; however, a vaccine will likely not be available until at least sometime in 2021.[31]

This statement was ultimately prophetic, as immunity did wane over time. But what we didn't anticipate was the impact of new coronavirus variants and how they would change the duration of immune protection from previous infection. There was little evidence that new SARS-CoV-2 variants were resulting in diminished protection among those recently infected and who recovered through the summer and early fall of 2020. That would come with the appearance of the Alpha, Beta, and Gamma variants late in 2020.[32] Meanwhile, we had increasing evidence that the new mRNA vaccines could be game-changing, since those vaccinated in the early studies developed substantial antibody immunity with two doses. At this point, I revised my professional opinion of the concept of lockdowns. Recall that Mark and I published an op-ed in the *Washington Post* in March 2020, arguing against widespread lockdowns because they would never be feasible for extended periods of up to two or three years.[33] But now we were within months of a possible vaccine Hail Mary, and trying to get as many previously uninfected people as possible safely to vaccination was an appropriate public health recommendation. Well, it turned out not to be. At this point in the pandemic, most people were unwilling or unable to distance themselves from others to remain uninfected.

This experience points out the changing nature of information. The statement our group made in the CIDRAP Viewpoint in April 2020 was on target. The op-ed piece Mark and I wrote in the *Post* was on target. And the reasoning for urging people to avoid infection in the fall of 2020 and get vaccinated in early 2021 was on target from a theoretical standpoint. From a practical implementation standpoint, I acknowledge it suffered miserably. The facts in this situation changed with time, and this had to be acknowledged.

Failure to Address Misinformation Credibly and Empathetically

Peter divides what we would term "misinformation" into three categories: demonstrable falsehoods, opinions you disagree with, and "factoids that are technically true but literally misleading."[34] The first two are self-explanatory. An example of the third would be stating that the trials of the initial Covid vaccines that were granted FDA Emergency Use Authorization did not yield statistically significant evidence of reducing the death rate from the virus.[35] This was technically true, because that kind of data would have required much longer clinical trials, but it was literally misleading because there was good evidence suggesting that the vaccines would cut down on death and serious disease.

Peter's guidance is that it is important to distinguish among the three categories but only regard the first as genuine misinformation that should be refuted with all available evidence. "Rebut only #1," he advises. "Feel free to explain why you think #2 is mistaken and #3 is misleading, but stop claiming they're misinformation."[36]

He also warns that rebutting mis- and disinformation can have the unintended consequence of giving it greater exposure. But if it is worth rebutting despite that possibility, go all in and repeat the point you're rebutting. "Rebuttals of unmentioned falsehoods rarely work. If you truly want to change people's minds, I think you can't just tell them the truth; you have to respond explicitly to the falsehood they currently believe." At the same time, it's important to validate their reasons for believing it, which is crucial in demonstrating empathy and collective goals. Peter suggests we might even acknowledge that the burden of proof is on the one who wants to change opinions and say something to the effect of "What you believe makes logical sense, whereas the actual truth is counterintuitive."[37]

Part of that validation is listening to the other viewpoint, something officials and experts often find difficult. But if you do that, and search for any points of agreement that can be built on, the people you are trying to convince will be much more likely to listen to your own decision process and take into consideration the point you are making. Peter says, "The more you know about how your audience came to believe what they believe, and the more you know about how some people have moved off that view to yours, the better able you will be to construct the pathway."[38]

Now, there will always be a cohort we'll never be able to reach, because they are too far gone, are outrageous in their assertions, or have their own agenda that doesn't square with objective truth or reality. Every time there is a serious infectious disease outbreak, for instance, we hear about how it was designed and manufactured to target certain racial or ethnic groups and spare others. Another example is the false association between early childhood vaccination and autism. There is absolutely no evidence of causation, as has been borne out in numerous studies and an entire book by Dr. Peter Hotez, professor of pediatrics and molecular virology and microbiology at Baylor College of Medicine and codirector of the Texas Children's Hospital Center for Vaccine Development, which explained what actually caused his daughter Rachel's autism and why it could not have been a vaccine.[39] More recently, Peter wrote an important book on the rise of anti-science, which underscores many of the trends that worry us.

Short of what some in our business refer to as the out-and-out crazies, though, there are opportunities to reach out with good communication. In early March 2020, I was asked to appear on *The Joe Rogan Experience* podcast to explain where we were in the emerging pandemic. I had never heard of Mr. Rogan, but I was told that he tended toward contrarian, conservative, or libertarian points of view and that he could be aggressive in his pushback against guests with whom he didn't agree. I try to be completely apolitical in all aspects of my work, so I wasn't keen on appearing there. Still, some of our younger staff members urged me to accept, saying he had millions of listeners, and we could get our message across to a large group we wouldn't normally reach through CIDRAP's news releases or any of our other avenues of communication. So, I agreed, and flew out to Los Angeles for the show, which dropped on March 10.[40] Based on the data coming in from all over the world, and my developing understanding of how this virus was behaving more like influenza than like previous coronaviruses, on the show I predicted that despite what WHO was *not saying*, not only did this certainly qualify as a pandemic, but it would likely last at least eighteen months and would cause hundreds of thousands of deaths in the United States and many more worldwide, and despite yearslong warnings from people like me and my colleagues, we were completely unprepared.

Throughout the lengthy program, Rogan was hospitable, cordial, respectful, and open to everything I had to say. I'd like to think he recognized authenticity when he encountered it. On the other hand, the pushback I received from much of the public and a sizable cohort of my professional colleagues was surprisingly vociferous, once again along the lines of *Who are you to scare the hell out of everyone?* When I was asked back on the Rogan show almost two years later,[41] we disagreed on several points. Fortunately, I had a CIDRAP staff member with me then who fact-checked my assertions in real time while we were on the air. Presented with the proof, Rogan eventually conceded that I was correct. At the time, this reaction gave me hope.

About a month after my first appearance on *Rogan* was when I did the interview with security expert Peter Bergen for CNN in which I asserted that the data and trends suggested at least 800,000 US deaths, based on the facts that the virus was obviously highly infectious, the global response was fragmented, we didn't have proven antivirals and other medical countermeasures in time, there was inadequate personal protective equipment for first responders and healthcare workers, and a vaccine would be at least a year away, and even once it was available, we didn't know how effective it would be and how quickly production could be ramped up. This was simply a mathematical calculation, I explained, based on the numbers of those infected and the rates of serious illness, hospitalization, ICU admissions, and deaths. These facts notwithstanding, I was pilloried as a scaremonger who just wanted to make headlines.

In April 2021, I came under attack when Nate Silver, the prominent statistician, writer, and then proprietor of the widely read FiveThirtyEight website, called me out on social media for being "relentlessly negative," posting, "He's spinning every fact in the most negative possible light. He's downplaying the effectiveness of vaccines. It all sounds terrifying."[42] Silver took me to task for saying on *Meet the Press* that month that "we're just at the beginning of this surge," that the virus "can actually evade the immune protection from the vaccine or from having previously been infected," that "I'm even more worried about what's coming down the pike over the next several years," and that "this B.1.1.7 variant is a brand-new ballgame."[43] In point of fact, while the B.1.1.7 variant didn't hit all regions of the country with dramatically increased case numbers,[44] subsequent variants—including Delta[45] and Omicron[46]—did.

Silver's comments reflected a serious problem we experienced throughout the pandemic: media stars opining on Covid science but not knowing what they were talking about. I do not consider myself above criticism, and I'll be among the first to admit when I'm wrong about something. Yet I never saw an admission from Silver—or any of the others like him who took me to task—when my predictions proved accurate; nor was there any attempt to correct the public record. I take no joy in being right about something as horrific as the loss of so many of our loved ones. I only wish that when misinformation is spread about me or my colleagues, maligning our efforts to prepare the public for what is ahead of us, at some point the record would be revisited, perhaps helping people see more clearly whom to trust in the future.

Politicization

We will deal with politics and policy in a subsequent chapter, but suffice it here to say that when Mark and I were writing *Deadliest Enemy,* our 2017 book on the worldwide challenges of infectious diseases, the one thing we didn't consider or feel a need to deal with was the politization of public health. Recalling how parents in the 1950s lined up and clamored for the opportunity to get their children vaccinated against polio, we never imagined that in a subsequent era, personal decisions on vaccination against a pandemic virus would become an extension of one's political views or an assertion of personal freedom. On the political front, whether an individual believed SARS-CoV-2 originated in the animal population or in a Wuhan research lab was often governed by his or her political affiliation rather than any real consideration of the evidence. And going back to our *fake consensus* issue, neither side was particularly eager to even recognize the other in its communication.

For example, those on the left roundly condemned President Trump's large, tightly packed indoor political rallies, which possibly led to the death of former presidential candidate Herman Cain,[47] among others. Trump supporters and those on the right instead focused on the mass protests following the police murder in Minneapolis of George Floyd, asking why we in the scientific community weren't decrying those clusters of people. In fact, the Minnesota Department of Health conducted intensified disease surveillance after the Floyd protests and found no significant increase in cases at these outdoor events.[48] The two sets of events were different, not only politically, but scientifically: indoor versus outdoor air. Clearly, though, we didn't communicate that properly. Later in the pandemic, when we were dealing with variants, we did see notable transmission at large outdoor concert venues when people stood packed together for hours.[49] Again, I emphasize that the communication cannot remain static; it must be updated as the specifics change over time.

Politicization can sometimes lead public health into no-win propositions. With the limited supply of Covid vaccine during the initial rollout beginning December 2020, authorities were criticized for not making minority and low-income populations priority candidates over healthcare workers, first responders, and highly placed government officials, since those low-income and minority groups constituted most of the people performing essential jobs like bus and delivery drivers, facility cleaners, store clerks, and factory workers and were therefore in great danger from the virus. There is validity to this argument. Yet I can well imagine that if these populations had been prioritized for vaccination, the same authorities would have been accused of turning this defenseless cohort into guinea pigs to make sure the vaccine wasn't harmful for the more affluent sector of society. This would seem just as valid an argument, given our shameful history of neglect and unethical medical response to people of color. After the forty-year Tuskegee Study beginning in 1932, in which the US Public Health Service studied Black men with syphilis, misinforming them about what it was doing and not actually treating them,[50] the government had an enormous task in getting people of color to trust it. Is it any wonder that many expressed vaccine hesitancy?

Yet, as Sandra Quinn notes, "When you talked to Black barbers and beauticians and other community members about [what happened] early in the pandemic...it was like the golden oldies of conspiracy theories brought up to date....These people were being hit hard. They have a history of being distrustful of medicine and government, and a past that supports that. But what we also saw, with ongoing dialogue, was people moving from distrust to...'This is happening. How do we protect ourselves? How do we protect our families?'" And, she adds, "Small-business owners, such as barbers and beauticians, that's their income. If they don't work, they don't make money. So they were eager to figure out how to protect themselves and their clients as best as possible."[51] Sandra also points out that barber and beauty shops were social meeting points, and so a good place for trusted communication about health to be shared. Sometimes a trusted messenger can be a peer, which, of course, can be a good or a bad thing, depending on what they say. This again underscores the importance of having communication channels in place so that more people are getting accurate information—and the importance of avoiding not just the mistake of politicization but also the other errors that Peter has identified.

THE MEDIA

Mainstream media posed its own challenges in reporting reliability. Every news organization in the world converted reporters who had no expertise or even experience in health reporting, let alone infectious disease, and assigned them to the Covid beat. Their editors and producers had little to no expertise in this area, either. The necessity for media organizations to quickly begin reporting on a news story that had highly technical aspects for which they had little experience contributed significantly to the problems of good journalistic communication during the Covid pandemic. I saw firsthand, day after day, how the vast majority of reporters, through no fault of their own, didn't have the expertise to critically evaluate the information they used to tell their stories. In addition, op-ed writers or commentators who had no expertise in scientific research methods took it upon themselves to provide supposedly expert commentary on issues like aerobiology and respiratory protection, vaccine immunology and virus seasonality. No news organizations were immune from this challenge, even three years into the pandemic.

One of the key lessons we must learn from the Covid pandemic to better prepare for the communication challenges of the Big One is how we can identify and promote the most reliable sources covering all aspects of the pandemic. After almost fifty years in public health, I have come to count on specific journalists who get it right, story after story. Although tempted to list them here, Mark and I are loath to inadvertently leave anyone out, so instead we offer a description of the characteristics of those we most respect: To the best of their ability, they vet their sources carefully; they check the facts presented by their sources; they explain things in a way that is accessible to readers or viewers; and they provide corrections if information is later found not to be accurate.

As we've suggested, the reporters covering a pandemic can only be as informed as the critical information provided by subject matter experts. Without it they can't do their stories. Sadly, many so-called experts actually weren't experts so much as what we call "epistemic trespassers."

In a 2019 issue of the journal *Mind,* published by Oxford University Press's Academy Research Forum, Arizona State University philosophy professor Nathan Ballantyne defines "epistemic trespassers" as "thinkers who have competence or expertise to make good judgments in one field, but move to another field where they lack competence — and pass judgment nevertheless," and cautions, "We should doubt that trespassers are reliable judges in fields where they are outsiders."[52]

During Covid, there were plenty of such trespassers populating the airwaves, internet, and print media, and so long as they had impressive-sounding credentials in *something,* their opinions were accepted by editors and producers as authoritative. We saw this on subjects such as respiratory protection, epidemiology (e.g., seasonality of the virus), vaccine effectiveness, and disease transmission modeling. There will be no easy way to guard against a repeat of this when the Big One hits, so we — both the public health community and individual citizens — should be aware of this media tendency so that we can at least be in a position to evaluate what these people say and advocate.

Even more problematic are those who are credentialed in associated (or adjacent) fields, like Dr. Peter Minekal in our scenario, who consistently offer wrong or misleading information. Once someone made the media circuit, he or she often became a talking head. I put myself in that category, so I don't automatically conclude a talking head isn't a part of an effort to educate the public about the facts of the pandemic. Throughout the Covid pandemic, the media picked and chose who would be interviewed for a story, be it in print or on TV or radio. The journalists, editors, and producers behind these decisions were the "king- and queen-makers." Those guests who became established voices were often interviewed again and again, almost like a self-fulfilling appointment conferring unique expertise. There was great reluctance among many talking heads or experts to take issue with points made by others because we were all taking so much flak, but it is vital that we speak out. We should be held accountable for the reliability of our message, and we must be willing to correct the record when we've erred. In our roles as public health and medical leaders, we have to call out our own, no matter who they are, when their personal beliefs conflict with critical public health recommendations.

On his MSNBC program on February 4, 2022, author and journalist Mehdi Hasan did just that. He interviewed Dr. Monica Gandhi, professor of medicine at the University of California, San Francisco, who had been a frequent and popular media presence during the Covid pandemic. After reviewing some of her pronouncements, such as that the virus wasn't a significant risk for children, Hasan drilled down. "Dr. Gandhi, one of the big criticisms of you is that you make these sweeping and very optimistic predictions, and they just don't pan out. I just want to run through some of them tonight so you can respond."[53]

He then proceeded to list a series of declarations and recommendations she'd made up to that point in the pandemic.

"You said last February, a month before India's horrific second wave began, killing hundreds of thousands, if not millions, quote: 'India now has herd immunity.'"

"I was very wrong about that," Gandhi conceded.

His enumeration continued with her denial of the significance of the evolution of viral variants; her reasons why boosters would not be needed; and her assertions that California would achieve herd immunity, that we would be in the endgame by the previous October, that the Omicron variant would not swamp hospitals in vaccinated areas, and so on.

"At what point, Dr. Gandhi, do you say, 'Maybe I should stop making predictions about a pandemic that I keep getting dangerously wrong'?"

Hasan ended up with at least a qualified apology from Gandhi, but his research, probing, and journalistic due diligence remained the exception to a dispiriting rule, even from institutions previously known for scrupulous research. Within days, Gandhi was back on the circuit, making the same types of erroneous statements.

I often say that throughout the Covid-19 pandemic, I saw some of the best reporting coming from the *New York Times*. I also saw some of the worst reporting coming from the *New York Times*. Nor was that august newspaper of record alone in those extremes. Too often, as we've suggested in this chapter's scenario, major news organizations perpetuated misinformation by continuing to rely on people who sounded authoritative as "reputable" sources without keeping track of whether what they said was accurate over time. Seldom did I see any media outlet report that something one of their quoted experts said was wrong or issue a correction, so there was no way for the reader or viewer to know from one week or month to the next what was accurate and what was not. If I'm a consumer of news, how am I able to distinguish between the good, the bad, and the ugly of all the stories and commentaries? With all the reporting on the failures of public health, politicians, and others in their work during Covid, I'm not aware of a single journalism-related professional organization or major media company that has considered a comprehensive review of what went right and what went wrong in media coverage of the pandemic. Yet such a review is critically needed if we're to be prepared for the Big One.

SOCIAL MEDIA

The internet and social media are the Wild West of modern communication, where the problems with traditional media are multiplied exponentially, since *everyone* can be a reporter; and as emergency medicine doctor Erin Thomas reflects in our scenario, *everyone [has] their own reality.*

We weren't prepared for the role social media would play in a pandemic before Covid-19. Obviously, there was no such technology in 1918, and between the influenza pandemic of 2009 and the emergence of SARS-CoV-2 in 2020, so much about the way we use the internet has evolved that we didn't have many lessons on how to use it as a communication tool. Even the way we access it has changed. Consider that the first iPhone came out in 2007, with Android models released the next year.[54] Just eighteen years later, they have become virtual extensions of ourselves, the primary way many of us interact with the world around us: texting, emailing, engaging social networks, following others, and posting on countless apps and platforms. Public health needs to stay current and present on all the major platforms, understanding the demographics of users of each, so that information can be tailored to the audience as well as the medium.

First, the good news. Social media provided a sense of community during the pandemic, when many would otherwise have been socially, as well as physically, isolated. People derived emotional, mental, and physical health benefits from keeping up with friends and family. They also shared useful tips on where vaccines were available and helped others make appointments to get vaccinated. While we can't know what developments the future will bring in technology and applications, we can anticipate that similar interactions will be beneficial during the Big One.

But social media has proven itself a decidedly mixed blessing.

In 2015, a user on Tumblr posted a picture of a striped dress and asked if it was white and gold or blue and black. The photo was reposted by others on Facebook and Twitter, and within thirty minutes, the controversy went viral.[55] It was covered on morning television and countless print and online media sources. It was a photograph of an actual dress, confirmed by the retailer who sold it to be blue and black, yet the world could not reach consensus as to the colors.

We mention this because, while it may be frivolous good fun when the subject matter is fashion, this type of arguing against facts can have devastating consequences when it involves critical information amid the spread of a deadly virus, particularly when there is so much unknown about the pathogen that even true experts on the subject don't have the answers, as the president bemoans in the scenario opening this chapter.

During the 2003 SARS outbreak, David J. Rothkopf described this phenomenon in an article in the *Washington Post*. His complete credentials are too numerous to list, but he is an internationally recognized journalist, entrepreneur, policy adviser, and national security and intelligence expert, and was then member of the Health

Advisory Board of the Johns Hopkins Bloomberg School of Public Health.

Rothkopf wrote,

> SARS is the story of not one epidemic but two, and the second epidemic, the one that has largely escaped the headlines, has implications that are far greater than the disease itself. That is because it is not the viral epidemic but rather an "information epidemic"... the information epidemic—or "infodemic"—has made the public health crisis harder to control and contain....
>
> An infodemic is not the rapid spread of simple news via the media, nor is it simply the rumor mill on steroids. Rather, as with SARS, it is a complex phenomenon caused by the interaction of mainstream media, specialist media and internet sites; and "informal" media, which is to say wireless phones, text messaging, pagers, faxes and e-mail, all transmitting some combination of fact, rumor, interpretation and propaganda. It can be rendered more difficult to understand by multiple languages, cultures and attitudes toward the free and open flow of information. It involves consumers of information ranging from officials to private citizens who have varying abilities to see the whole information picture, varying degrees of sophistication about what to do with the information they have, little opportunity to authenticate data before acting on it, and little if any training in understanding or controlling the rapidly changing information picture.[56]

We saw this happen with Covid. On February 2, 2020, more than a month before it would characterize Covid-19 as a pandemic, WHO warned in its *Novel Coronavirus (2019-nCoV) Situation Report—13*, "The 2019-nCoV outbreak and response has been accompanied by a massive 'infodemic'—an over-abundance of information—some accurate and some not—that makes it hard for people to find trustworthy sources and reliable guidance when they need it." In the same report, it assured readers, "WHO technical risk communication and social media teams have been working closely to track and respond to myths and rumours."[57] WHO and the UK government rolled out a global campaign—"Stop the Spread"—to raise public awareness of the risks of believing and spreading misinformation about Covid-19. The campaign included a call to action that sounded a lot like what people said about the virus's spread: "Let's flatten the infodemic curve."[58]

So, what can we do about the infodemic? People will continue to post and repost whatever they believe is valid, and we cannot expect that to stop. If you felt overwhelmed, welcome to the club. Even with my experience, it was difficult not to drown in the latest posts, and I would often get media requests to comment on posts the same day they went online. I don't know how anyone without professional training or experience in epidemiology, public health, risk communication, virology, vaccinology, and public policy could navigate it all; yet to prepare for the Big One, we must anticipate an even more voluminous social media impact.

I navigated social media by identifying a few of what I considered the best and most reliable sources and following only these individuals. For the Big One, I would propose a similar approach: that I and other pandemic responders share our trusted social media colleague list with whoever might find it helpful. For Covid, I closely followed reporters Helen Branswell, John Burn-Murdoch, and Sanjay Gupta, along with eight other social media regulars.

At the top of the list—and really in a category of his own—was Eric Topol, MD, director of the Scripps Research Translational Institute. Eric consistently provided the most comprehensive and visionary commentary throughout the pandemic. The others I followed closely were Peter Hotez, MD, PhD, dean of the National School of Tropical Medicine and professor of pediatrics and molecular virology and microbiology at the Baylor College of Medicine, whom we mentioned earlier; Andy Slavitt, former acting administrator of the Centers for Medicare & Medicaid Services and temporary senior adviser to the Covid-19 response coordinator in the Biden administration; Ezekiel Emanuel, vice provost for global initiatives at the University of Pennsylvania; Trevor Bedford, PhD, professor at Seattle's Fred Hutchinson Cancer Center; Florian Krammer, PhD, professor of vaccinology at the Ichan School of Medicine at Mount Sinai in New

York; Kristian Andersen, PhD, Department of Immunology and Microbiology at Scripps Research; and Trisha Greenhalgh, OBE, professor of primary healthcare at Oxford. These respected colleagues were a professional lifeline for me.

Unfortunately, many other commentators regularly posted what I believe was dangerous mis- and disinformation about Covid vaccines, drug treatments, and the risk of infection. I have no doubt the mis- and disinformation from these individuals was directly responsible for many people refusing vaccination or appropriate treatment, leading to unnecessary deaths. Heading up this list are Alex Berenson,[59] Robert F. Kennedy Jr., and Joseph Mercola,[60] though there are plenty more we could name.

As in so many other areas of life, confirmation bias—the tendency to believe information that agrees with what an individual already thinks about something—can creep into the equation and influence the way people process or react to what they read. And the longer a pandemic drags on, the more the fatigue factor sets in; people just don't want to have to hear or think any more about it.

We saw an initial drive for tech companies to police the information posted on their platforms, but that seems to have either fallen by the wayside or proven too difficult to execute effectively. As the debate between those arguing that Meta (which owns Facebook, Instagram, and WhatsApp), X (formerly Twitter), YouTube, etc., have a responsibility to at least flag, if not remove, information that is patently false and those insisting that doing so constitutes suppressing free speech continues, we see initiatives like Meta's Transparency Center potentially raise as much frustration among legitimate users as false information does. The task is challenging: Blocking or removing bad information is not an exact science, as I learned when two of my podcasts were removed from YouTube because I'd said things containing phrases its filters used in banning false content— never mind that the context of my use of those key words was to explain why the assertions they represented were wrong!

In concert with the Stop the Spread drive, WHO set up a web page with information on "how to report misinformation online"[61] and links to several major social media platforms. However, once you follow a link to a specific platform and report a concern, it is not clear what, if anything, will happen regarding the errant—or malicious—post.

As with other forms of media, it comes down to being an informed and responsible consumer of information: Think critically about what you see, hear, and read.

While we encourage individual critical thinking, we can't avoid the responsibility of leadership—in public health, science and medicine, politics—to provide accurate information in real time. We would hope that official sources with reliable communication be active in all social media spaces, but until we know we can count on them for accurate and up-to-date reporting, including making clear what is still unknown, that may not help us much.

At a minimum, however, reliable public health messaging should be present on *all* popular platforms, and that messaging needs to be written and produced in a way that will reach the divergent audiences on each one. For example, during Covid, tweets were often worded so as to reach scientific and medical professional audiences, while TikTok videos were produced with music and engaging production to appeal to a younger audience. Managing social media presence is a specialty in itself, and we need experts to advise and help craft the messaging.

Back in 2009, a case study by Jennifer Aaker and Victoria Chang from the Stanford University Graduate School of Business, entitled "Obama and the Power of Social Media and Technology," traced how a little-known first-term senator from Illinois and his innovative team understood and marshaled more than fifteen social networks to get his identity known and his message out. The case study concludes, "They had a very talented candidate who was a great communicator and they had a campaign philosophy that matched and mirrored very well with the Internet—openness, inclusiveness, self-organizing, grassroots."[62] We need to run our pandemic communications campaign with the same efficiency and sophistication.

Of course, none of this works if you don't have the "product" to

go with it, and in our case, that means up-to-date scientific and epidemiologic information, again combined with the message of humility
in acknowledging that we are dealing with an evolving situation and
constantly getting new information. The communications environment has changed, and social media has become more sophisticated
since Barack Obama's first presidential campaign. We must recognize that public health cannot afford to cede any of this ground.
And we must keep in mind that if a pandemic lasts three or more
years, what was a hot platform for communication in the early days
may not be nearly as effective several years in. Technology, culture,
and user expectations evolve nearly as fast as viruses, which leads us
to our next subject.

ARTIFICIAL INTELLIGENCE, SYNTHETIC MEDIA, AND THE THREAT OF DEEPFAKES

Artificial intelligence (AI) — the development and use of com-
puter systems that can perform tasks considered to require human
intelligence — is popping up in applications from virtual assistants
to self-driving cars. With generative AI — i.e., the kind that can be
used to produce text content, images, etc. — anyone can quickly
and easily produce content on a given subject. There's great potential for positive uses for these applications, but the technology also
raises profound concerns regarding the creation and dissemina-
tion of disinformation.

While the companies behind this technology build in so-called
guardrails designed to prevent users from creating malicious content, disinformation, and hate speech, it is not difficult for moti-
vated users to work around them. Prompted effectively, generative
AI technology can produce hundreds of articles and blog posts full
of disinformation, including fake testimonials, inaccurate or mis-
represented data, and lists of apparently legitimate but actually
made-up references. Sadly, research has shown it can be very difficult to discern whether the output is real or fake.[63]

Even more disturbing than text is other forms of what is called synthetic media—video, voice, or images generated in part or fully by AI. Deepfake videos can replace a person's voice or face, making it look as though someone is delivering whatever message the creator wishes. We have already seen manufactured voice and video used maliciously in everything from political campaigns and sophisticated bank fraud to phone scams in which people receive a call that sounds like it's from a loved one who is in trouble and needs money.

POSSIBLE FIXES

Who decides what is information, misinformation, or disinformation? That question is at the heart of the challenge to vet the avalanche of scientific papers and preprints and media stories, including mainstream and social media, that will occur with the Big One. What is truth, and who owns the expertise to determine it?

Historically, the most trustworthy sources of health information were healthcare providers and government agencies. But even these sources became less reliable as they were unable to expertly evaluate and summarize Covid-related scientific information as it exploded in volume and reliability. For example, between January 1 and June 30, 2020, there were approximately 24,000 scientific publications about the SARS-CoV-2, treatment, and public health prevention activities, including the use of nonpharmaceutical interventions.[64] This included research articles, letters, editorials, notes, and reviews. The medical and public health communities were simply unable to vet and summarize this explosion in information in a way the public could meaningfully use.

Is there a compromise to be had between withholding scientific studies until they are peer-reviewed and rock-solid in their evidence, and keeping the public and political leaders informed on relevant and timely data? Is there an effective way to combat misinformation that spreads rapidly through the various media?

We propose the creation of an international epidemic and pandemic information intelligence crisis command operation—a "validation bureau" with the personnel, expertise, and resources to evaluate mainstream and social media statements, preprints, publications, and other preliminary findings and announcements, and sufficient capacity and funding to provide as close to real-time, expert review as possible, with no more than a twenty-four-hour turnaround. All information reviewed and a summary of its accuracy would be published on the command center's website, organized in such a way that the public can search the database by subject, date of publication, source of the information, and lay summaries of scientific findings.

It shouldn't be an organization like WHO or the CDC, because they can be perceived as having a vested interest in what is put forth, and so might not be considered credible by a portion of the population. It must be independent, unbiased, and have "no dog in the fight." We suggest establishing such a center well before the Big One, with emergency plans to substantially ramp up activities with the pandemic onset. Further, we recommend following the CEPI model: a global partnership comprising philanthropic, public, private, and civil society organizations; encompassing countries, the European Commission, multiple foundations, and academic institutions.

If such a center had been in operation during Covid-19, the subject of "masking" might not have become such a political hot button, for example, and people would have understood that N95s were protective, while everything else was of limited protection. Scientific information in a pandemic is too vital not to have such an independent and trustworthy means of evaluation.

In any war, reliable information is among the most valuable weapons. Our war against deadly microbes is no exception.

TAKEAWAYS

1. Crisis communication is larger and more diverse than any one organization or institution, and we must recognize that it has to change over time.

2. Official communication needs to balance authority with humility, presenting the most up-to-date information along with the acknowledgment that we don't yet know as much as we will, that the science is steadily evolving, and that information and guidance may change as more data are compiled and more is learned.

3. Compassion and empathy are critical in conveying the idea that there will be difficult and challenging times ahead but we will get through the crisis together if we cooperate and show kindness and consideration toward one another. A reassuring tone from a leader that everything that can be done is being done will go a long way in establishing public trust and compliance.

4. Establishing and using communication channels that flow in both directions, from high-level public health, science, and medical leadership to local, trusted messengers in all communities, *before* the Big One will do a lot to ensure that accurate, effective information is shared in time to make a difference.

5. Realistic bad news is easier for people to accept than unrealistic optimism or minimization of the risk. The truth may not set us free, but it will give us realistic parameters and goals.

6. Public health officials should admit when they are wrong. Otherwise, they forfeit their credibility. And when they admit they're wrong, they should explain why they were wrong.

7. Minority opinions should not be ignored. The public should understand that all reasonable points of view have been

considered and are continually under review as the crisis unfolds.

8. Public health considerations are not the only ones in a pandemic. Society must continue to function as well as possible. Officials must communicate that all decisions are based on trade-offs and that all options have been carefully weighed and balanced.

9. The media should keep track of what its quoted experts say and should correct statements that are wrong. They should likewise keep track of who has been consistently correct. If there were an independent, unbiased organization to vet statements and track records of media-designated experts, as well as information available on the internet, that could be an additional monitor of reliable communication.

CHAPTER SIX

SURVEILLANCE

The reason for collecting, analyzing, and disseminating information on a disease is to control that disease. Collection and analysis should not be allowed to consume resources if action does not follow.
          —Dr. William H. Foege, CDC director, 1977–83

[Pandemic Month 12]

As the SARS-3 pandemic passed the one-year mark, it continued to be the number one news story throughout the world. Although the first vaccine—an mRNA platform as was used to fight the Covid-19 pandemic—had been developed and was available within six months of the initial outbreaks, several disturbing developments mirrored what happened with the mRNA Covid vaccines. First, it took far longer than hoped to scale up manufacture and distribution, which was particularly frustrating for those in the United States who wondered why the necessary infrastructure hadn't been established after Covid. Second, even when widely available, the vaccine faced not only hesitancy but outright hostility among certain groups and population segments.

Perhaps the most disconcerting parallel to Covid was the growing incidence of infection among the vaccinated, who were getting sick four or more months after vaccination. After a drop in cases around six months, a new much more severe wave was playing out. A new variant had emerged.

And the pandemic fallout reached into other critical areas. Many links in the worldwide supply chain had collapsed, as China and India reported massive worker shortages due to illness and death at major manufacturing centers. The same was true for the crews of the more than 60,000 freighters that plied the world's oceans, as well as for the dockworkers who loaded and unloaded them. And even if a ship could make the journey to the United States from Asia and be unloaded in port, there was a major shortage of truck drivers and rail employees to move the supplies to their final destinations. Together, these factors created worldwide shortages of products and their components, food shipments, and pharmaceuticals. Persons with diabetes struggled to find insulin, and prices shot up. Healthcare providers found themselves having to ration immunosuppressive drugs, antibiotics for patients with infections, and many anticancer agents. Supermarkets periodically shut down as food shortages surged. Nursing homes and prisons were experiencing huge outbreaks with each new wave of infections, with the elderly the most vulnerable. Stock markets around the globe had already lost close to half their value.

Because of inadequate international surveillance and case reporting, no one knew how many people had died from the virus, but it was clearly in the tens of millions. Nor was the United States spared. As with Covid-19, it had close to the highest mortality rate among high-income nations.

Contrary to what many of the people being interviewed on television and quoted in the media were optimistically saying about the virus once again beginning to trend downward, some experts were warning that the darkest days were still ahead—comparing the current situation to the moment just before the Covid-19 surge when the Omicron variant became dominant. Already the world had seen outbreaks come in waves, as variants hit people whose immunity from the first SARS-3 mRNA vaccine had waned and those who had refused to get vaccinated altogether or had no access to the vaccine. But with SARS-3, the rates of mortality and serious illness were far higher than with Covid. One public health official was quoted as saying, "This is like Omicron on steroids!"

Though there was controversy about the total number of cases, one thing was clear: Scores of smaller hospitals had to close their doors to new patients because so many of their doctors, nurses, and support staff were out sick. One major hospital system reported more than 1,000 workers out at one time.

Home SARS-3 tests had been approved for use for close to ten months, but like everything else, their availability was dependent on supply chains, and with repeated surges in infections, there weren't enough to meet the demand. And, as with Covid, SARS-3 testing was tricky. If you didn't test at the right time, you might not test positive even though you were both infected and infectious. What's more, those who did test at home did not report results to any state or federal agency, making it impossible to get an understanding of the true spread and scope of the disease.

The best data US public health officials could get regarding vaccine effectiveness and who was at greatest risk of severe illness or death were from Canada, Israel, and the United Kingdom, due to their national healthcare systems, which allowed for rapid linkages between vaccination status and illness history. American officials then had to try to extrapolate from those figures what might be happening in the US population. The optics also weren't favorable. By now, many politicians, celebrities, sports figures, and prominent media personalities had died of the virus. A New York Times *reporter who had been covering the pandemic had spent three weeks in the intensive care unit. He had texted from his hospital bed, "I guess this really is THE BIG ONE we've all been worried about."*

Dr. Tamara Goldfield sat at the end of the long table in the CDC director's Incident Command Center and faced the wall-size panel, divided into screens that currently displayed the images of state epidemiologists and lab directors from around the country. The CDC's principal deputy director, Dr. Jacinda Trowbridge, sat to her left. Other senior CDC staff sat around the table. Repeatedly, Goldfield heard from the state officials that they couldn't keep up with the challenges the pandemic imposed on them. They were exhausted and overrun. Already understaffed before the pandemic, after a year of playing catchup, they were beyond running on fumes, as many of their staff were out sick, had left because of burnout, or, most searing of all, had died or become at least temporarily unable to work because of the effects of the virus. Almost all said they needed help and financial support. Goldfield silently wished she had more to offer them beyond the prospect of a new and better vaccine in the works.

Those were the general complaints. Specific issues varied by state. Some were getting active pushback from their governors, who were succumbing to public pressure not to shut down schools and businesses or require vaccination

and threatening to replace all the leaders of their state health departments.
Others were receiving demands from teachers' unions to close schools. Others
were still fighting the masking wars. Everyone needed answers. Should they
shut down businesses temporarily to try to flatten the infection curve and keep
hospitals functioning? Was it more harmful to close schools or keep them
open? Offline, she'd gotten disheartening calls from several state health offi-
cers who told her they wanted to comply with CDC recommendations, but state
and local leaders forbade anything that made it look like they were "locking
down" and threatening businesses.

The question of vaccination was even more fraught. The governors of
several red states were actively urging people not *to get vaccinated, even as*
local hospitals and clinics were completely overrun with SARS-3 cases. In her
own experience, she'd seen that while many people were eager to receive vacci-
nation, it seemed that just as many were resisting, saying that if the vaccine
didn't stop the spread of SARS-3, or prevent you from contracting it even if
you were vaccinated, what was the point?

Goldfield conceded the shortcomings of the mRNA vaccine, in terms of
both its waning immunity over time and its inability to prevent people from
getting infected and infecting others. But she pointed out that after an admit-
tedly slow start and insufficient funding, the efforts to develop and scale up a
more effective and robust vaccine were well underway, working with inject-
able viral particle and recombinant protein platforms; an adenovirus vector
type that could be orally inhaled, targeting the mucosa so that the virus would
be largely neutralized before it reached the lungs; and a live attenuated vac-
cine. All four candidates were already nearing completion of Phase 2 trials
and looking very promising. FDA Emergency Use Authorization could possi-
bly come within weeks or a few months for one or more of them. There were still
questions about how they would work over time, as long-term protection
hadn't been ascertained, but she was hopeful at least one would prove to be a
game changer.

In spite of this potential step forward, Goldfield herself was tired, frus-
trated, and at the end of her emotional rope, though she tried hard not to let
the stress show, which often only made it worse. She needed data in order to be
effective. The White House was demanding it so they would know what to do,
and the only way the CDC could compile the necessary statistics and numbers
was to get them from the states. But when she articulated this, as calmly as she

could manage, the state officials said they were in the same boat. Hospitals, clinics, and doctors' offices were just as overwhelmed, facing the same challenges they were, and were not forthcoming with the numbers needed for active and up-to-date surveillance. How else would they get a handle on this public health disaster?

I HAVE COME TO believe that "surveillance"—a word first used in reference to public health some seventy years ago—is no longer the best term to describe a vital aspect of our efforts to identify and manage threats from infectious disease. It makes it sound like we're talking about spying on people, when what medical or public health surveillance is really about is gathering and processing the intelligence required in our ongoing war against potentially dangerous or deadly microbes. There has been an effort recently to refer to this as "disease data collection and analysis." However, the word "surveillance" is still widely used, so we will use it here.

Regardless of the term, when we think of medical surveillance in a public health context, we tend to focus on two complementary activities. First, in its broadest sense, it is the day-to-day tracking of important health events in our communities, whether they are due to an infectious agent or another cause, like cancer or a death in childbirth. The earliest use of public health surveillance is still the most common reason we conduct it today: to learn about everyday infectious diseases in our communities, like those that cause foodborne illness, are transmitted by mosquitoes, or could be prevented by a vaccine. Second, we attempt to identify emerging disease threats around the world, similar to what we described CIDRAP doing in the Prologue, when Lisa Schnirring contacted me in December 2019 about the mysterious respiratory disease in China. Ideally, the practice serves as an early warning system and usually starts with questions like, What looks suspicious? Is anything unusual going on with any wild or domestic animals? Are farmers or agricultural workers who tend pigs and birds presenting with any new respiratory complaints? Do we suddenly see cases of measles in an area of our community? Are there any unexplained pneumonias cropping up? We are also concerned with whether Ebola or some other hemorrhagic disease has materialized from the wilds of Africa, though as we've explained in earlier chapters, it doesn't take a crystal ball to know that the next pandemic will be a highly infectious respiratory illness.

It is helpful to distinguish between two basic types of surveillance. *Sentinel surveillance,* as defined by the CDC, is "reporting of health events by health professionals who are selected to represent a geographic area or specific reporting group."[1] *Syndromic surveillance* is the early warning system we mentioned earlier, such as when the CDC collects data from hospital emergency departments on specific conditions or symptoms related to infectious disease. Internationally, WHO requires reporting of certain infectious diseases, including smallpox (which we hope never to see again in nature), human influenza caused by a new subtype, and SARS. This kind of robust surveillance is critical to give us a heads-up and early start on any new infectious agent of epidemic or pandemic potential. But just as critical, perhaps even more so, is what we do once we identify such an agent and find ourselves in the midst of an epidemic or pandemic. We have to be able to track it in close to real time to determine where the disease — most certainly a virus — is spreading; who has it and what their common profile is, if any; what the patterns of spread are; how many people are sick enough that they end up in the hospital or ICU; and what vaccines, drugs, or other medical interventions are, or may be, effective in dealing with it.

During Covid, we were bombarded with numbers from around the United States and the world that showed we were dealing with a disaster. But in too many situations, we didn't know what specifically was happening in our communities and other geographical locations, because we had only limited information at that level. We needed a level of granularity in data that was often not available. It was like a great forest fire: From a distance, you can see the smoke, but you know exactly where and how large it is only when you get much closer.

We have to improve our ability to see through the smoke. In this chapter, we will explore how we got where we are now, what we can learn from the Covid-19 pandemic, and what we must do to develop and maintain an efficient, effective disease surveillance system that will provide actionable data to inform public health decisions—not only when the Big One hits, but also as we are challenged with health threats every day.

A LONG HISTORY

Disease surveillance is not a new concept. In fact, we have references to Hippocrates in ancient Greece promoting the idea of recording and collecting disease details in the Athenian population to help determine what measures should be taken.[2] During the Middle Ages, cities and towns throughout Western Europe monitored various illnesses, leading to regulations regarding gutter and water pollution, food handling, and burial of the dead.[3] By the early fourteenth century, Venetian officials were authorized to board incoming ships to make sure no one aboard was carrying plague.[4]

In late seventeenth-century England, John Graunt published *Natural and Political Observations Made upon the Bills of Mortality* and set forth the basic principles of disease surveillance, including death rates and patterns of disease spread.[5] Graunt is frequently referred to as both the first demographer and the first epidemiologist.[6] Across the Atlantic, in 1741, Rhode Island passed a law requiring tavern keepers to report various contagious diseases, including cholera, smallpox, and yellow fever.[7] It took more than 125 years for the first public health agency to be established in the United States, in 1866—in New York City—but by the turn of the century, as we noted in Chapter Three, forty states and some local jurisdictions had followed suit, and Congress had established the forerunner of the Public Health Service to collect medical data that would inform decisions on quarantines and other mitigation measures.[8]

Dr. William Farr, superintendent of the statistical department of the Registrar General Office of England and Wales from 1839 through 1879, amassed a wide range of what he called vital statistics that he reported to health and municipal authorities based on which they

could make decisions and act.[9] Interestingly, Farr had originally sub-
scribed to the miasma theory of cholera until statistical analysis,
including data from John Snow's investigations, convinced him otherwise.[10] Just as Snow is regarded as the father of epidemiology, Farr is
widely recognized as the father of surveillance.[11]

In Italy, compulsory reporting of infectious diseases began in 1881,
and in 1890 in Great Britain.[12] By 1901, every state in the United States
required reporting of certain diseases, such as cholera, smallpox, and
tuberculosis.[13]

The surveillance and reporting trend continued throughout the
first half of the twentieth century, buoyed by the polio epidemic
of 1916 and the Great Influenza of 1918. Fifty years later, at the
twenty-first World Health Assembly in Geneva, Dr. Alexander Langmuir of the CDC presented a working paper on global disease surveillance that he had developed in consultation with physicians and
epidemiologists from around the world[14] on how to achieve the three
goals he would go on to stress in 1972 in a seminal lecture delivered
to the Massachusetts Medical Society. Many in our field credit this
lecture as the beginning of modern public health surveillance. He
defined it as the systematic collection of significant disease data; the
consolidation and organization of the data; and the timely dissemi-
nation of the data to the relevant authorities and those who needed
to act on it. He stated that the term "epidemiologic surveillance"
implied "the responsibility of following up to see that effective action
has been taken."[15]

Alex Langmuir was already something of a living legend by this
time, the chief epidemiologist for the Communicable Disease Center and its successor agency, the Center for Disease Control (and, as
of 1992, the Centers for Disease Control and Prevention). In 1949,
he established what became the CDC's Epidemic Intelligence Service (EIS), a corps of young disease detectives who would go anywhere in the world to investigate and try to solve mysterious
outbreaks, with the intent to bring them under control. He sold Congress on the idea by stressing its importance as a mechanism of
national defense against the possibility of biological warfare. He
coined the term "shoe leather epidemiology,"[16] suggesting there was

no substitute for getting out in the field to get the facts and see the patients firsthand. That phrase remains the unofficial EIS motto. Today, many thousands of physicians, veterinarians, epidemiologists, statisticians, and other healthcare specialists have gone through the EIS program, at headquarters in Atlanta and stationed at public health agencies throughout the United States and many locations around the world. Alex personally trained at least the first five hundred.

As another living legend, former CDC director and one of Mark's and my personal heroes, Dr. William H. "Bill" Foege wrote in a 1996 article in the *American Journal of Epidemiology,* "What was [Langmuir's] message? That epidemiology is the basic science of all public health *and* that there are scientific rules for the optimal application of epidemiologic methods.... He revolutionized the way epidemiology was used in public health practice, first in the United States and then throughout the world."[17]

This all has deep personal resonance and meaning to me as an epidemiologist for nearly the past half century. Among my most treasured possessions is a first edition of *Vital Statistics,* a memorial volume of selections from William Farr's writings and reports, published in 1885, two years after Farr's death, by the Sanitary Institute of Great Britain. The book has its own peculiar history. It was purchased from a rare book dealer in London in 1940 by Alex Langmuir. In October 1966, Alex gave the book to his protégé and close colleague, D.A. Henderson, then chief of the CDC's viral disease surveillance programs, as D.A. was leaving for Geneva to head up the World Health Organization's smallpox eradication effort (in which Bill Foege played a major role). Alex inscribed the book,

To D.A.

A memento of 11 wonderful years together promoting and extending the principles of the father of surveillance.

A.D.L.

Smallpox was one of the great scourges of history, and what D.A. and his expert and dedicated worldwide team accomplished in the next decade was truly remarkable: the first time a deadly infectious disease had ever been completely wiped out. D.A. and Alex were both inspirations and cherished mentors to me, as they were to so many others in our field. So, it was incredibly moving and meaningful to me when, in February 1999, D.A. presented the book to me with the inscription:

To Mike,

Alex Langmuir, as I'm sure you know, credited William Farr as being his inspiration for the creation of the concept of surveillance which has proved to be so critical in the practice of public health and so pivotal to the successful eradication of smallpox and polio.

On my departure from CDC for Geneva, in October 1966, Alex presented to me his personal copy of the memorial volume of Farr's papers. There is a note at the front and you will see on page xvii, Alex's handwritten comment, "no mention Snow?"

It seemed to me that at some point in time, it would be appropriate for me, in turn, to pass this on to someone whom I felt best exhibited those attributes which Alex prized so much—a rather inquisitive mind; an avid interest in surveillance and in field epidemiology; an ability to distill from the dates and "all other relevant sources" (as ADL would express it) salient conclusions; and the necessary courage to act upon one's convictions even when the data were less complete than one would like but when the public health was at stake.

I can think of no more worthy recipient than Mike Osterholm. Herewith, then, the passing of the torch as you embark on a new adventure.

D.A.

I never have received, or will receive, another such honor in my life. And in keeping with the tradition set by these two great heroes of public health, at some future date I will either pass the volume on to a worthy successor or bequeath it to a major medical archive.

THE BUILT-IN CHALLENGES

The reasons for conducting daily public health surveillance should be pretty clear by now: to detect outbreaks, epidemics, and pandemics; to describe the potential for, and burden of, a particular disease; to monitor changes in disease-causing pathogens; to identify areas for research; to provide data for the development and prioritization of policies, programs, laws, and mandates; to keep the public informed; and to evaluate prevention and control strategies. In a pandemic, we need these data in order to understand how the disease is spread and what the specific risk factors are, such as comorbidities, age or ethnicity vulnerabilities, social behaviors, etc.; to estimate the impact of the disease on a community, such as burden to hospitals and other healthcare facilities, to aid in resource allocation; and to evaluate the effectiveness of strategies to prevent or combat the disease, both nonpharmaceutical interventions and drugs and vaccines.

Currently, there are various means for public health authorities to collect, compile, and aggregate disease surveillance data, constrained by federal and state laws and regulations, legal precedence, and financial resources.

One challenge we face is that the primary surveillance system used by public health agencies in the United States is routinely composed of paper or electronic reports from healthcare providers to local or state public health departments. This can be simple and inexpensive or complicated and time-consuming, depending on the number of surveillance staff in a department and the existing systems for sharing these data. When electronic reporting systems are in place and agreed-upon protocols for what information needs to be reported (such as name, address, contact information, and laboratory data) are standardized, the system can be timely and effective. Unfortunately, for many health departments, case reporting information is often far from complete, and it varies greatly in quality. Health agencies often must reach out to providers to elicit additional data that were not included in the original report. For example, a report without a case home address or contact information makes it impossible to determine the location of the case. On the other hand, while these specifics are necessary for state and local health department investigations, for surveillance purposes, names and other personally identifying data are not needed by the CDC or any agency managing a pandemic.

The question becomes, How do we manage an enhanced system like this? Such a proactive approach to follow-up can yield more complete information, but it is time- and labor-intensive, and it requires trained personnel. In addition, when potentially hundreds or more reports come into a health department every day, adequate electronic data management becomes critical for timely review to detect outbreaks. Sadly, the surveillance system we just described, with incomplete case information supplied on paper forms, is the norm for much of the United States, and it will not change without adequate resources being directed toward the effort.

Another built-in challenge to obtaining the kind of data needed to make accurate, time-sensitive decisions affecting pandemic response in the United States is that legally, the kind of data collection we're talking about is reserved to the individual states, not the federal government. This means that the CDC cannot come in and compel data and statistics from the local frontline treatment centers, even in a national emergency. Under the Commerce Clause and general welfare provisions of the Constitution, Congress has broad power to protect the "general welfare," but the police powers, including those required to ensure public health, are reserved to the states. The CDC gets involved when invited by a state, or when a disease or medical condition has interstate implications. However, the Commerce Clause is grounded in protecting and promoting the national economy, so the federal government would have to show a close connection between the nation's economic interests and the need for medical data from states.

The states themselves decide individually which infectious diseases are reportable, and what is on that list and how it is reported are highly variable. The reporting entity could be a lab, a clinical office, a hospital, or a long-term-care facility. Sometimes the data are compiled by the state; other times this work is done locally, in which case a local health department is responsible for conducting case investigations and actions, then forwarding the data to the state health department. In still other states, data and reports come directly to the state health department from the individual health-care providers—physicians, hospitals, local laboratories. There is no universal reporting structure.

Contributing to the confusion, there are various systems used by tribal territories and the more than 3,300 county and city health departments.[18] This gives you an idea of how many different kinds of entities we are dealing with. And to complicate things further, much of the communication between local, state, and national authorities is still done using antiquated technology. During Covid, some health departments were still using fax machines.

Not only do the data come in via outdated and variable pathways, but because states get to decide what is a public health priority and report data accordingly, the information collected and made available to the CDC is often inconsistent, making situational awareness on a large scale difficult, if not impossible. In the midst of the Covid-19 pandemic, for example, Florida prohibited reporting to the CDC, not wishing to be held up publicly for how many cases were occurring in the state.[19] And not every state, municipality, or health-care system is even collecting the same information, so we see large knowledge gaps. Louisiana, for example, reported Covid data to the CDC, but its health department personnel often had to obtain race and ethnicity data on Covid patients by cross-referencing the information they received from local providers with Department of Motor Vehicles, Medicaid, and other source records in an effort to identify who was at most risk of infection and severe outcome.[20]

As with so many aspects of public health and healthcare, disease surveillance in the United States is not reaching much of the population. Those who do not have health insurance, Medicare, or Medicaid, or who cannot seek medical treatment for any number of reasons (unhoused or undocumented individuals, for example), or who live in areas where one would have to travel a great distance to reach a clinic or hospital are underrepresented, at least until a dramatic outbreak hits.

These are among the factors that have "left a mess for those of us tracking this novel [SARS-CoV-2] virus," wrote health and science journalist Betsy Ladyzhets in a March 23, 2022, article on the FiveThirtyEight website. "For two years we've tried to make sense of COVID-19 trends with metrics that were fundamentally impaired by our chronically decentralized and underfunded public health system....If the country doesn't want to repeat its mistakes, it will have to take radically different actions the next time a health crisis hits."[21]

WHAT WE NEED

We need our public health surveillance to be more like the National Weather Service. Every day of the year, the weather is different — not only different from the day before, but also different from the weather that day in other locations. Most of the time, only local experts need to know what's happening. But if portions of Tornado Alley suddenly light up, or there is going to be flooding in California or a hurricane in Florida or on the Gulf Coast, everyone in the affected area needs to become aware and take appropriate actions. Behind local National Weather Service forecasts are extensive nationwide ground radar and weather stations, as well as satellite systems, that provide huge amounts of data in real or near-real time. These data are analyzed twenty-four hours a day using enormous computing power. National Weather Service meteorologists and climate experts incorporate this information into comprehensive statistical models to make predictions with ever-increasing accuracy for weather conditions seven or more days into the future, and long-term temperature and precipitation predictions for at least a month.

We need this type of ongoing and accurate system to monitor infectious diseases around the world, with numerous sources and types of actionable data available 24/7/365. Additionally, we need similar substantial investment in technology to process and parse the data. And just as meteorologists get even more data, more quickly, in times of severe weather emergencies, we need to be able to scale up and access additional resources when the Big One hits. Right now, in the face of a pandemic, it becomes obvious that we have a major disease crisis going on, but beyond that, we need to understand with accuracy how many people are sick, the individual characteristics of the people who get infected, exactly where they are, and what's happening to them to determine what types of public health recommendations can be brought to bear to reduce transmission. So, how do we get the information we need?

Local and State Resources—Our First Line of Defense

Our first line of defense is our local and state health departments. The beginning of our thought experiment in the first chapter of this book demonstrated why: Somali community health worker Jamilah Shamshi was the first healthcare professional to observe the new disease, and when she saw how quickly case numbers were rising and how serious the illness was, she escalated up the chain of command. Similarly, we can expect that the first harbinger of a new pandemic (or any type of outbreak) in the United States will be a cluster of cases presenting to a local clinic or hospital emergency department. This means our routine disease surveillance resources need to be strong at the local and state levels—including all communities.

When I was the state epidemiologist in Minnesota from 1984 to 1999, infectious disease surveillance was an all-consuming pursuit. Our team at the Minnesota Department of Health published numerous papers in medical journals about effective surveillance and came to appreciate how we suffer from our hodgepodge of reporting. I am also painfully familiar with how even the most robust local systems are underfunded, understaffed, and often lacking the human resources and technological tools they need to be effective.

The Council of State and Territorial Epidemiologists (CSTE), a group I have belonged to since 1976, serving as president from 1993 to 1994, has the best front-row seat at what it takes to achieve the surveillance systems of the future that we need. They are the disease-reporting experts, and I strongly agree with the CSTE position that all case reports, ideally delivered by efficient electronic means, should first be reported to the state and local agencies, as they are now, allowing review of the case information at the local level to identify incomplete reports that need immediate follow-up and to conduct a daily assessment that might indicate that a possible outbreak is emerging. It should then be transmitted to the CDC, after first being de-identified—i.e., providing vital demographic data without individual names.

This problem of timely, thorough reporting will not be solved until we get necessary resources to state and local public health agencies and enable nationwide electronic case reporting. I have yet to see any comprehensive and realistic attempt to define the resource needs in order to establish this kind of national system, so I hope those in authority start listening to CSTE as the expert surveillance resource group. Otherwise, we are doomed to repeat the same surveillance challenges we experienced during Covid with the Big One, with even greater consequences.

Efficient and Comprehensive Data Collection and Reporting

In the United States, we can draw an analogy to the air traffic control system. Both systems work to an extent and are operated by dedicated, hardworking people, but they suffer from equipment and capacity far outdated for the current demand. Additionally, with both systems, there is no safe margin for error, and we have seen the number of near misses and other mishaps grow over the past few years. Admittedly, both the system for tracking and directing commercial airliners and the system for tracking and analyzing infectious disease will be costly to upgrade, but the cost of not upgrading will be far higher in terms of lives put in peril and our nation's ability to maintain efficient commerce and lifestyle.

Clearly, the current agglomeration of disease surveillance systems is not serving us very effectively. For the future, we need improvements in our surveillance methodology—both during a pandemic and for routine situational awareness—as well as in our technology. Our experience with Covid gives us some clear examples of what we need to consider in both areas.

During the Covid pandemic, we saw that the countries that tend to do best at surveillance once a serious infectious disease is circulating are those with centralized healthcare delivery and payment systems, like Canada, the United Kingdom, and Israel. They have the means to know the number and demographics of those who are sick (without individual names), how many hospitalizations there are on a given day, and how many deaths; and they have access to actual lab data. "The U.K.'s Health Security Agency has been a particularly popular data provider when new variants emerge, with its regular reports showing the connections between variants and changes in transmission, hospitalization rates and vaccine effectiveness," Ladyzhets wrote in her piece entitled "The U.S. Still Doesn't Know How to Track a Pandemic."[22]

Contrast this with the United States, where in April 2020, only thirteen states were regularly reporting current Covid-19 hospitalizations, according to *The Atlantic*'s highly regarded Covid Tracking Project.[23] The number of states that were reporting increased over the course of the pandemic, but the numerous "data holes" in the early months meant that we were never able to get ahead of the situation.

In the earliest days of the pandemic, when an effective Covid vaccine seemed like a distant hope, public health, medical, and political leaders considered any plausible means to slow virus transmission and reduce illnesses and deaths. For many, the answer was contact tracing. As the CDC notes in a publication (updated online in September 2022), "Contact tracing has been an important tool for epidemiologists who are trying to stop a disease outbreak," stating further, "**Contact tracing** has been used by CDC and its partners as part of successful strategies to help stop disease epidemics like small-pox, Ebola, *and now the COVID-19 pandemic*" (italics added).[24] In addition, the CDC has used contact tracing as one method of conducting surveillance, which allows scientists to gather data about infections and how they may spread.

This lies at the center of a critical debate we encountered during Covid, and one that we will face again going forward. As we consider what tools will be useful when we are confronted by the Big One, it is vital that we understand the limitations of contact tracing.

This public health tool has played an important role in controlling the spread of SARS and MERS, leading many to conclude that contact tracing could be successful in controlling Covid. By late March 2020, the enthusiasm for state and local health departments to develop extensive tracing programs was unbridled. And again, I found myself trying to swim up a scientific Niagara Falls. I had major concerns about how effective these efforts could be in slowing down SARS-CoV-2 community transmission, based on one simple fact: What we'd learned so far about the virus was that its transmission was different from that of SARS and MERS, as well as measles, TB, Ebola, and STIs, where most cases were not infectious until becoming symptomatic.

Of note, in the more than twenty-five years I have participated in international and national preparedness planning for an influenza pandemic, no practical, long-term contact tracing plan was included. The reasons will become clear as we examine the challenges of contact tracing for Covid.

There are four factors that determine the potential to identify and interrupt the transmission of infectious disease cases with contact tracing. First, is the virus or bacteria transmitted via the airborne route? If so, there are likely to be many individuals who are exposed to the case in public and private settings without knowing where the exposure came from. Think of all the people you don't know whom you share air with on an average day. In contrast, a sexually transmitted infection leaves little doubt, for most people, about how the exposure happened.

Second, what is the time from exposure to becoming infectious? For example, with measles, symptoms appear seven to fourteen days after exposure, with a rash appearing three to five days after the first symptoms.[25] The individual is typically infectious from ten to twenty days after exposure,[26] allowing for contacts to be identified and provided information on isolation with the earliest symptom onset and prior to being infectious.

Third, what is the frequency of asymptomatic transmission? Can transmission occur prior to first symptoms or among those who are infected but don't experience any clinical illness?

And fourth, is there a drug treatment or vaccine that can prevent infection or minimize clinical disease if caught in time? For example, smallpox had an average incubation of ten to fourteen days. However, vaccination within four days following exposure can be effective in preventing infection and transmission to others.[27] Vaccination four to seven days after exposure did not prevent infection but prevented severe symptoms and death.[28]

By late February 2020, I believed the epidemiologic evidence on Covid was clear: An increasing percentage of SARS-CoV-2 transmission was occurring from infected but not yet (or never) clinically ill cases. While the contribution of SARS and MERS asymptomatic cases to transmission has not been well characterized, it appeared substantially less than we were experiencing with Covid. Contact tracing for Covid therefore could not be justified based on those two diseases.

In the early months of 2020, it was estimated that the average incubation period for Covid was around five to seven days.[29] With each subsequent variant, the incubation period decreased. By the time Omicron emerged, we estimated that the incubation period was as short as three days,[30] and up to 50 percent of infectious cases were asymptomatic.[31] That meant that throughout the Covid pandemic, the epidemiologic cards were stacked against contact tracing as an effective method for reducing community transmission. You can't trace contacts of asymptomatic cases because such cases never come to detection. And with an incubation of only two to four days, by the time someone became symptomatic and got a positive test result, it was often four to six days after illness onset and five to seven days after that person was infectious. Even with an immediate interview of cases by a public health worker and the identification of possible contacts, many of the contacts were well into their infections.

Efforts in January and February 2020 by public health officials in China, South Korea, and Singapore suggested that contact tracing and mandatory quarantine did help limit the size of their initial local Covid outbreaks. The actual impact of contact tracing in these countries was difficult to measure, however, as it was accompanied by other control measures taken at the individual and community level, such as stay-at-home orders and banning public gatherings. In the United States, as I soon saw firsthand, there was little consideration of the four factors that determine whether contact tracing may be effective in interrupting transmission to a next generation of cases.

On April 9, 2020, I was asked to attend an online meeting sponsored by the Rockefeller Foundation on developing a national Covid testing action plan and the launch of a Covid community healthcare corps for testing and contact tracing. After listening for forty minutes to a discussion about how the activities detailed in the plan could have a major impact on the course of the pandemic, I raised all the issues that I believe made a national contact tracing initiative a poor use of resources.

One of the other participants in the online meeting was Paul Romer, 2018 Nobel laureate in economics and professor at New York University. As I was laying out my concerns about a national contact tracing program, he abruptly interrupted me and launched into an attack on my small thinking about controlling the pandemic. He asserted that I had no business even participating in this online meeting, railing against me to the point where a member of the foundation had to intervene; the following day, four people who attended the meeting called to express their apologies for the Romer outburst. If nothing else, Paul's comments as a Nobel laureate, but someone with no expertise in this area, reinforced the fact that the contact tracing public policy train had left the station.

Other highly visible organizations joined in the chorus strongly supporting contact tracing as a key component of local and state health departments' efforts to contain Covid. One was Resolve to Save Lives,[32] headed by Dr. Tom Frieden, former CDC director in the Obama administration.

Given the pro–contact tracing climate, I brought together a group of "the old band" in early May to author a CIDRAP Viewpoint on the subject.[33] The group included two retired state epidemiologists from New York and Colorado, a former senior official in the CDC sexually transmitted disease program, and former senior Minnesota Department of Health officials who had also worked in contact tracing. We tried to provide a balanced and objective review of the utility of contact tracing for Covid and concluded the report with four pressing issues.

First, data are needed to do the following: determine if there is sufficient benefit to justify the cost of widespread contact tracing; define what level of exposure is truly significant for Covid and requires self-quarantine of contacts; evaluate compliance with self-quarantine, particularly when the source is not readily apparent to the contact; define benchmarks for success; and clarify the usefulness of digital tools for contact tracing, particularly addressing privacy concerns.

Second, guidance should be provided on how to prevent potential adverse impacts of widespread contact tracing for Covid.

Third, technology standards are needed for using digital tools for contact tracing.

And fourth, public health officials need to define parameters for using contact tracing during different phases of the Covid pandemic, including determining the endgame for when this effort can be de-escalated or completely demobilized.

Our Viewpoint was downloaded more than five thousand times in just the first week it was published. Despite the sense that contact tracing had the momentum, by May 2020, a growing number of public health professionals were pressing the same issues we raised in

the Viewpoint. Nonetheless, tens of thousands of new contact tracing staff were being hired or reassigned from other positions in public health agencies nationwide.[34] This was all driven by federal
direction and support.

Still, the initial public health effort with contact tracing was
mind-boggling to me. For example, the Minnesota Department of
Health employed more than 1,200 people in that effort. Other health
departments across the country followed suit. With Minnesota's population of 5.5 million, 1.7 percent of the US population, you can better understand the thousands of people and resources dedicated to
this one effort to limit cases on a national basis. The money spent on
contact tracing was in the many millions of dollars.

On May 26, 2020, the CDC released its interim guidance on
developing a Covid investigation and contact tracing plan, recognizing some of the same issues we identified, stating, "When a jurisdiction does not have the capacity to investigate a majority of its new
Covid cases... [it] should consider suspending or scaling down contact tracing."[35]

A series of studies conducted by the CDC, state and local health
departments, and academic research groups in the United States
and Europe attempted to determine the effectiveness of contact
tracing. The results were mixed at best. Notably, two studies led by
different groups of epidemiologists and statisticians at the CDC
came up with conflicting conclusions.[36] The fact that the CDC could
produce two separate studies on Covid contact tracing effectiveness
with such divergent results gives a sense of how soft the data were for
making a case for these programs. Yet, as we noted, a CDC training
publication on the topic suggested it was a successful strategy as late
as May 2022.[37]

One other more comprehensive and better-designed study evaluated the proportion of secondary infections captured by contact
tracing in Geneva, Switzerland, from June 2020 to February 2022.
The authors confidently concluded, "Contact tracing alone did not
detect sufficient secondary infections to reduce the spread of
SARS-CoV-2."[38]

By 2021, public health agencies nationwide started to pull back

from contact tracing due to the workforce and financial requirements. Contact tracing of individuals stopped in many localities,
and efforts were concentrated on institutional transmission such as
in schools and long-term-care and correctional facilities.[39]

Then, in January 2022, the final nail was hammered into the
contact tracing coffin. The Association of State and Territorial
Health Officials, the Council of State and Territorial Epidemiologists, the National Association of County and City Health Officials,
the Big Cities Health Coalition, and the Association of Public Health
Laboratories issued a joint statement supporting public health
departments that were moving away from universal case investigation and contact tracing.[40] They supported more strategic approaches
to outbreak investigation and targeted contact follow-up.

I hope we will not make contact tracing a centerpiece of any
future influenza or coronavirus pandemic response. Such an
approach may be helpful for controlling emerging coronavirus outbreaks like SARS and MERS, but it won't be for a highly transmissible airborne virus-caused outbreak with short incubation periods
and significant asymptomatic transmission. This is a critical lesson
for the Big One.

A more practical approach to data collection that has popular
support among the public health community is random sampling.
As Emory University biostatistician Natalie Dean wrote in a 2022 col-
umn for *Nature,* "Random sampling can answer those [important
risk-benefit] sorts of questions. As long as participants are selected
randomly, they will on average mimic characteristics of the wider
population."[41]

The United Kingdom's REACT (Real-time Assessment of Community Transmission) program followed this methodology.
Launched in April 2020 to track Covid infections, the program was
funded by the Department of Health and Social Care and administered by Imperial College London in collaboration with the Ipsos
MORI market research company.[42] Over a series of "rounds," REACT
collected test swabs and blood samples from 120,000 to 180,000 randomly selected volunteers to gauge the level of SARS-CoV-2 infection, detailed by age, ethnicity, occupation, geography, and other

indicators. In some rounds, participants were incentivized with shopping vouchers.[43] As Dr. Marc Lipsitch of Harvard told us, "Because it was a random sample, it was a real gold standard for how many people were actually infected at any given time, so they could get prevalence really directly in a way that's different from number of cases. It could compare the number of cases, but all the things that keep people from getting properly counted when you're counting cases were not an issue because it was a random sample, everybody had a test kit, and they weren't testing because they were sick, or because they weren't sick, or because they were traveling, but because they were asked to do it. . . . It was an amazing thing. I mean, the UK had its problems in Covid, but high-quality surveillance was not one of them."[44] Random sampling would also help balance the challenge of unreported at-home test results. Ironically, as home Covid tests became widely available during that pandemic, we in the United States actually had *less* actionable information, since almost none of these results were reported anywhere. Although initially launched to study Covid-19 disease prevalence, REACT is now being used to gather information on long Covid, as well as influenza, showing the utility of such a system even outside the context of a pandemic.

We have relied on programs like REACT and those from other countries with centralized healthcare systems to inform our vaccine and drug strategies and help us make informed decisions. But while using data from other countries is better than nothing, it does not really give us a picture of what is going on in our own country in real time, especially given how regionally diverse our country is. Some of the general surveillance methods currently in use, such as wastewater analysis and reporting the number of hospital beds occupied, are helpful, but they are nonspecific.

Wastewater surveillance is one public health accomplishment that is often overlooked for its long-term benefits. Some bacteria and viruses that infect us are shed in our feces and have been detected in wastewater dating back to the 1940s. In the 1940s, routine wastewater surveillance for polio gave public health officials an early warning when the virus was circulating in a community, before it was detected clinically.[45]

By the 1980s, molecular techniques were able to identify infectious agents without having to grow them from waste samples. Then, in the 1990s and early 2000s, PCR made it possible to detect the presence of very small amounts of bacteria and viruses or their particle debris,[46] but wastewater sampling using PCR was done on a limited basis only. Covid changed that. In February 2020, SARS-CoV-2 was detected in wastewater in Amersfoort, Netherlands, by Dutch officials, six days before the first cases of illness were diagnosed in that community, demonstrating that SARS-CoV-2 is excreted in the stool before one becomes clinically ill and by those with asymptomatic infections.[47] Wastewater sampling has also played a critical role in monitoring mpox and H5N1 influenza activity.

The current program is coordinated by the CDC through the National Wastewater Surveillance System (NWSS). By late 2024, it covered more than 1,500 wastewater sites, representing a population of more than 150 million, and continued to grow,[48] and included state, local, tribal, and territorial health departments, wastewater treatment plants, and public health, environmental, academic, and private laboratories.[49] I can envision a day when many more infectious disease agents will routinely be included in the testing protocol, and future technical advancements will further refine the sensitivity and specificity of the testing.

Also during Covid, the CDC's Genomic Surveillance Program requested that volunteering incoming passengers be screened through nasal swabs at major international airports, and individual international flights had their wastewater collected and analyzed to try to get a jump on emerging and incoming viruses.[50] But these efforts, far from complete, lacked the comprehensive information we needed in order to take action. They gave us a current picture of what was happening infection-wise in a community or county, but not the specifics on *who* or *how many* were infected. In this way, they can serve as an early warning system, giving us situational awareness, but only timely, consistent, universal, and detailed data reporting—who, where, what, when, all throughout the United States—gives us truly actionable information.

Clearly Defined and Updated Case Definition and Surveillance Objectives

Another issue that sounds basic but is vitally important is the case definition; that is, defining the symptoms, lab findings, and other relevant data that determine who has the disease. On the most basic level, this guides health authorities and frontline personnel to know what they're looking for and to be able to exclude other conditions. We don't want to be too specific, especially at the beginning, because we don't want to miss cases, so there is always a balance to be struck. For example, at the beginning of the Covid pandemic, the case definition focused on travelers who had recently been in China, which quickly became misleading when it turned out that asymptomatic transmission allowed the virus to spread around the world. Those coming into the country from northern Italy were just as likely to be infected as those from the region near Wuhan. Originally, too, loss of taste and smell were important indicators of SARS-CoV-2. But as the variants came along, those symptoms were not common, so it was critical to keep the case definition up to date to recognize who had it and who didn't.

Just as important as defining the disease itself is defining the objectives of surveillance. These may differ at the local versus federal level, or even from state to state, depending on population, resources, the current variant (if the virus in question mutates), whether a vaccine is available, and how widely a vaccination program is being accepted. Objectives might include trying to slow the spread, mitigation of severe demand stress on resources, understanding the genomics of variants and their changing impact on those infected, and trend reporting that identifies those most at risk and provides guidance on how to cut down that risk. There will be a natural tendency to want to respond to the public and media clamor for information that is interesting to that audience, such as a daily running tally of deaths, which is morbidly fascinating but takes valuable resources to gather and update each day. It makes more sense, from the standpoint of public health, to devote the limited resources of health departments around the country to focusing on getting data that will help answer questions like how well vaccines are working to prevent transmission, serious disease, hospitalization, and death.

The problem and challenge going forward is that the infrastructure simply does not exist in the United States to support the kind of surveillance that will help us weather the Big One. Covid showed that our systems were not capable of handling the overwhelming volume of data, which therefore often couldn't be gathered, processed, and analyzed quickly enough to be of actionable use. Various systems that had important data weren't integrated or interoperable with other systems and were difficult to modify in an emergency. When critical records did exist and could be reported, there was often no incentive for overworked gatherers and administrators to pass it along to higher-up authorities.

Some of the best case data on hospitalization and severe disease came from those covered by the Centers for Medicare & Medicaid Services (CMS), which has a set of Conditions of Participation for all the healthcare entities it deals with. In 2011, CMS instituted an Electronic Health Record Incentive Program. In 2018, that evolved into the Medicare Promoting Interoperability Program, "to encourage eligible professionals (EPs), eligible hospitals, and critical access hospitals (CAHs) to adopt, implement, upgrade, and demonstrate meaningful use of certified electronic health record technology (CEHRT)."[51] This includes a merit-based incentive payment system. And during Covid, payments from CMS were linked to reporting vital information.[52]

It's likely that the best way to obtain the necessary data is for the government to pay for it, which won't be cheap. But the payoff in cost savings from disease prevention and mitigation will be dramatic. As we've stressed, and will keep stressing, any money spent on collection and analysis of *actionable data* on public health threats, including infectious diseases of epidemic or pandemic potential, is an investment in national defense and homeland security; and anything we spend on our war against dangerous microbes is a lot less expensive than what we spend on the demonstrably lesser threat of conflict with human enemies. It would be possible during the Big One, for example, for CMS to require any of the data we feel are useful to be reported to the CDC and other relevant governmental agencies. Looking to the future, we ought to be able to figure out a complementary way to incentivize reporting for those whose care is reimbursed by other means.

Andy Slavitt, who served as the senior adviser on Covid response in the Biden White House, ran CMS for two years under President Barack Obama. He also has been a dear friend and mentor to me on many of the critical issues regarding the policy aspects of our Covid response. He commented to us, "There are a lot of different approaches. I'm not sure whether or not you have to pay for the information, or reimburse the cost [of organizing, compiling, and reporting], or make it part of participation in design.... Every hospital has a condition upon receiving payment from CMS that requires them to have their data available for real-time surveillance review on some limited number of things.... That's really your enforcement."[53] Other payers could follow a similar model. And we'll stress one more time, individual names of Medicare and Medicaid clients are not part of the needed information.

Assuming we achieve buy-in for reporting, what we must be able to do, in plain and simple terms, is connect the electrons of electronic medical records at the local level to the surveillance systems of the local, state, and federal public health agencies. The first mission would be what we are calling a Target Activity Profile, or TAP: define our surveillance objectives and figure out what activities will get us to the point of meeting our needs for information and data, sufficient to fully interpret what is gathered and be able to act on it. Andy Slavitt notes, "I think if you got five smart health data people, you'd probably get five really legitimately good and interesting answers," based on what it is you're trying to accomplish,[54] underscoring the need to involve a consensus of public health experts to define the TAP parameters that are appropriate and useful to meet specific objectives.

For example, Andy adds, "I'd make each vaccine vial electronically trackable....Knowing where they are, how they've been taken, etc., in real time would be very useful."[55] To be clear, this refers to tracking vaccine doses as inventory *prior to* their being injected into anyone. We are not talking about the sort of vaccine-dose-as-tracking-device approach popularly claimed by conspiracy theorists! Knowing how many doses were used in a specific area, and who received them, from where, would give us valuable information about how vaccination rates correlate to infection, helping us understand vaccine effectiveness. Knowing who received them can inform us on uptake rates in specific communities so we can reach out more effectively. And knowing where people received their vaccination can help public health officials understand if there are barriers related to factors ranging from trust in the source to logistics of location and hours of operation affecting whether people choose to get vaccinated.

Marc Lipsitch was the lead author on an article published in *Frontiers in Public Health* entitled "Infectious Disease Surveillance Needs for the United States: Lessons from Covid-19." The authors list critical decisions dependent on accurate and complete surveillance data, including "size of response needed; choice of community countermeasures; how to ensure adequate supply of hospital beds, ventilators, and personal protective equipment; schools and congregate setting policies...countermeasure deployment within a jurisdiction; [and] efforts to distribute and promote vaccination."[56]

"The good news is that a lot of the data [already] exists. It's not easily linked to everything about an individual, so the epidemiological kinds of questions are still going to require compiling some stuff together. But I don't think the issues are technical," Andy Slavitt observes. "There are [additional] things that would help, that would enable things better. But I think the issues have largely been, Are people willing to play nice together? For some questions, the sample's pretty damn good, like if you had, let's say, Labcorp, Quest Diagnostics, Walgreens, CVS, Rite Aid, Walmart...if you had that data, and you had it in real time, you'd know a ton."[57]

TAP can help us devise reasonable recommendations for the public. As Peter Sandman points out, if we give people metrics about what we propose to do when certain thresholds are reached, such as a percentage of the population infected or a certain percentage of hospital beds filled, they are more likely to take us at our word and react accordingly. They will understand, for instance, that schools might be closed for several weeks to flatten the curve, but they won't stay closed indefinitely. This will require near real-time reporting of data that is accurate and complete. What the TAP parameters should be for each public health action like school or business closing can be arrived at by the collaboration of working groups from various areas of disease surveillance and medical records management.

Most of the data TAP would require are available at the local or regional level, in large medical records systems like Epic and Cerner, used by many US hospitals and healthcare entities.[58] There needs to be a countrywide, coordinated effort to figure out how to connect these systems to a national, real-time database. For this, we propose what we're calling a national Medical ID Number for each individual that would have all of his or her significant medical information and data in one electronically accessible file. This would have the dual purpose of offering quick and appropriate treatment wherever we happen to be, as well as giving public health authorities immediate access to the kinds of surveillance data we've been talking about in this chapter.

A single Medical ID Number would give decision-making authorities a direct link from the point of each person's healthcare contact to the point of policy. Without using names or individual identities when not used to treat the individual's medical needs, these records could be mined for any number of actionable analyses, including easy random sampling. In normal, non-pandemic times, we would have near real-time data on risks from any disease on which we wanted to focus, vaccine and drug effectiveness, environmental influences on health, and a myriad of other issues. In addition to the benefits to the individual in healthcare delivery of having all health information in one easily accessible place, the uses and public health benefits of such a system would be profound. It would literally revolutionize the realm of disease surveillance.

Now, we would be naive if we didn't anticipate serious pushback to this proposal. There will be protests against invasion of privacy from people not wanting the government delving into our private lives and personal business. It could be just as politically polarizing as so much of the Covid-19 pandemic turned out to be. But in a way, this is akin to the punch line "Keep the government out of my Medicare!" Consider that, depending on our situation, we already consign our various records and sensitive information to the Social Security Administration; Medicare and Medicaid; the military or Veterans Administration; the State Department's Passport Agency; Global Entry and Trusted Traveler programs; the Department of Motor Vehicles; any HMO, hospital that treats us, or healthcare system we belong to; etc., etc., etc. Every time you use a frequent-buyer card or number at your local drugstore, you are contributing to their file of everything you've purchased, and how often. Then there are other organizations that have our sensitive records whether we like it or not, such as the credit reporting bureaus. And how many people regularly make personal information—including innumerable pictures of their face, which could be used by facial recognition software, and their voice—public or semipublic on social media without even thinking twice? The point is, there are safeguards in place that protect your personal information from unauthorized or improper use, such as HIPAA, the Health Insurance Portability and Accountability Act.

"You'd need a process of de-identification, which you have, to be able to have someone be able to query that data and do a study," says Andy. "It's not technically hard if it's being recorded the right way. The most important thing is that everybody complies.... It's not perfect, because I think you run into all those things in the US that are privacy values, all these sorts of things. But yes, technically, it should be possible."[59]

If such an individual Medical ID system were in place, supported by the proper resources and infrastructure, before the Big One or any other pandemic hit, the saving of lives and efficient allocation of hospital facilities, medicines, supplies, and equipment would be a game changer.

Marc Lipsitch, whom I rely on and trust as much as anyone on this subject, would like to see several surveillance systems in place for maximum effectiveness: something like TAP, random sampling, wastewater analysis, and—if it could be implemented—the Medical ID. "If you had several of these [tools in place] and you could triangulate them, you could get a pretty good real-time, or close to real-time, sense of what was going on. I think you need several things for each purpose, which, of course, would be expensive, *but pandemics are expensive!*"[60]

We would want to see data collection and analysis driven by public health experts rather than by outside consultants, as so often is the case with government agencies. I have found that consulting companies, sometimes referred to as "Beltway Bandits," generally don't have a deep knowledge of what surveillance data are needed, particularly on the community level. Also, bringing in consultants means that the agency is not developing critically needed resources and capabilities in-house.

The need to have true public health subject matter experts drive the solution was perfectly illustrated at a conference we attended at the National Academy of Medicine in Washington, DC, in 2023. A panelist from Omaha, Nebraska, asserted that hospitals should be responsible for overseeing case reporting, rather than public health agencies. This point was put into the context of reality when, not more than half an hour later, complaints were voiced that hospitals in Omaha wouldn't cooperate in sharing Covid case information with one another.

All the capabilities we describe should already be in place and working routinely *before* they are needed in an outbreak, epidemic, or pandemic. The same surveillance systems that we use every day should be what we can rely on in a pandemic, scaled up. It is similar to a municipality understanding its daily water demand and how much is available at any given time. If there is a major fire, the water needed to put it out will be coming from the same system used to deliver water throughout the community; officials figure out if they have to decrease water pressure in one area so they can direct more to the area of the fire.

There are some encouraging developments underway, for example, the CDC's Data Modernization Initiative, which it describes as, *"a multi-year, multi-billion-dollar effort to modernize data across the federal and state public health landscape.*...Our vision is to create one public health community that can engage robustly with healthcare, communicate meaningfully with the public, improve health equity, and have the means to protect and promote health."[61]

Effective and efficient surveillance is a key component of that aspiration.

TAKEAWAYS

1. Complete and efficient disease surveillance is critical for informed risk-benefit analysis of all public health administrative and leadership actions. Without efficient surveillance, we cannot understand the risk factors for various infectious diseases or the resources needed to combat them.

2. Our first line of defense is routine disease surveillance, and we must support resources at the local and state levels. CSTE scientific and technical leadership is a necessary component of building an effective surveillance system of the future.

3. We must look at disease surveillance as a seamless process from the point of healthcare delivery—in hospitals, clinics, doctors' offices, and home care visits—all the way up through the timely state and local health departments' reporting and delivery of non–personally identified case reports to one central repository at the federal level. We need to be able to collect information and test results electronically, and have the capacity to mine data quickly at any level of authority. Approaches and tools such as a TAP and a national Medical ID Number should be implemented to facilitate these efforts.

4. The Medical ID Number would have the dual benefit of informing public health officials of emerging infectious threats while

having an individual's complete medical history available to doctors and healthcare providers in any physician's office or emergency department, wherever that individual happens to be.

5. National contact tracing programs are not a practical or effective public health tool for responding to an emerging influenza- or coronavirus-caused pandemic.

6. States, localities, and individual hospitals and healthcare facilities should be incentivized to participate in electronic surveillance systems. If the state is included, the system will work, regardless of politics. Cases and related information will then show up automatically and be part of the national database.

7. Any new systems should be designed and administered by public health personnel themselves and not by consultants—the so-called Beltway Bandits. Additionally, they should be appropriate for everyday use so that those using them are familiar and comfortable with them when the Big One hits. The cost of their development will be paid back through utility over time—i.e., instead of wasting money, energy, or time on "one-and-done" resources developed for a specific public health emergency, we invest in systems that yield useful information continuously.

CHAPTER SEVEN

POLITICS AND POLICY

It's a very dark chapter to have anti-science, anti-vaccine activism adopted by elected leaders — by an extreme element of a major political party — and it is a killing machine.
— Dr. Peter Hotez, interview with CIDRAP News

[Pandemic Month 15]

As the president sat in his traditional seat in the middle of the long, oval table that dominated the Cabinet Room, he glanced from the participants gathered around him up to the portraits of Thomas Jefferson and Theodore Roosevelt directly across from him. His gaze then traveled unavoidably to those of George Washington and Harry Truman flanking the door on the wall to his right. He couldn't help wondering how each of his predecessors would have handled the SARS-3 crisis, and whether they would have approved of the job he was doing. More than that, he wished he could ask them for advice, particularly TR and Truman, who he felt would have been as no-nonsense and take-charge as any of them. At least he had made it a point to have regular cabinet meetings to evaluate, as objectively as possible, his administration's response.

He had followed Deann Morgan's suggestion to hold "Fireside Chats" on the pandemic, broadcasting them from the small study off the Oval Office. They had been lauded by much of the medical and public health establishment and by members of his own party. The opposition party and some in the media had labeled them opportunistic grandstanding, trying to create a crisis when there wasn't one and making the pandemic seem even worse than it was in an effort to get more people to accept vaccination. But any crisis that killed millions of people in the United States alone, and tens of millions more around the world, was certainly a crisis in his book, no matter what else any-one said about it.

At the White House, they were observing the strict protocol of everyone who worked there being tested twice a day, and everyone who came in for a meeting being tested before being allowed in. But the rest of the country could not be expected to follow such stringent measures. And infection control, the president well knew, was only one aspect of the challenge. The effects on the economy were never far from his mind.

The president called on Dr. Edward Winters, chairman of the Council of Economic Advisers. Winters, who had previously been a prominent labor economist at UCLA and MIT, had authored one of the most influential books in the field. "Well, Ed, how are we doing on your front?" It was the way he started nearly all these meetings, knowing the economy was nearly always campaign issue number one. With the next presidential election less than a month away, and seats in both houses of Congress up for grabs, the pressure was high, even if there was little he could do in the waning months of his sec-ond term, short of starting a war, to push the needle dramatically in either direction.

"As far as we can tell so far, sir, and you know an economist is never going to give you a straightforward answer," Winters began, "we think we've done a pretty good job of evaluating real need and balancing that against expenditures. It hasn't been perfect—you might say we're trying to paint this plane as we fly it—but, unlike what happened with Covid, as we came to understand, we haven't been just throwing money at the problem without some stress tests for who in the population actually needs the help, so as to try to avoid an inflationary spiral from too much liquidity in the marketplace. Just as importantly, I think, our efforts to subsidize businesses, especially small businesses, to keep workers on the payroll has been successful in terms of making the pandemic less economically disruptive, and it will make it much easier to get back to business as usual once the pandemic is over, whenever that turns out to be."

"I agree," said Secretary of Labor Arthur Eggleston. "If all the jobs the subsidies supported had to be refilled—either with workers rehired or new

staff trained—it would cause a massive dislocation that would delay economic resurgence even further."

The president turned to HHS Secretary Sulbarry. "Anne, how are things going on your side?"

"I'd like to let General Russell speak to that," she said. "I'm sure he wasn't eager to be asked to take charge of procurement and logistics at this stage of his career, but let me say, I can't think of anyone who would have done a better job." Brigadier General Caleb M. Russell had recently replaced Ronald Engler, a longtime political operative and colleague of the president who had been brought in and tasked with SARS-3 coordination, procurement, and distribution but had become a target of media pundits and some of the opposing party's more vociferous critics. With the election polling this close, the president and his advisers couldn't take a chance. Ron Engler's head had to publicly roll.

"Russ," the president said, nodding to Russell, "the floor is yours."

"I think Secretary Sulbarry is giving me too much credit," Russell replied. "All I tried to do was implement a program that was devised above my pay grade. But to address your question, sir, the rapid development, manufacture, and distribution of the first mRNA vaccine six months into the pandemic has had, honestly, mixed success so far. I understand from Secretary Sulbarry that less than thirty-five percent of the high-risk population has been vaccinated to date. We have successfully forward-deployed vaccine, but the hesitancy to get it is clearly carrying over from the Covid pandemic. That said, I'm pleased to report that the central coordination, procurement, and allocation of personal protective equipment, drugs, supplies, and medical devices like ventilators have made a huge difference in terms of equalizing access throughout the country, avoiding the issue of having governors and state health departments having to compete with one another, and preventing vendors and suppliers from jacking up prices. As you know, we haven't always had enough of everything we needed, especially in the early months, but this system at least gave state officials some confidence and peace of mind that they were all being treated equally."

The president turned back to Sulbarry. "Anne, where are we on the new vaccine front?"

"We're still cautiously optimistic about the novel mucosal vaccine the NIH has been developing in cooperation with some of the pharmaceutical

companies," Sulbarry replied. "The funding you were able to secure for research beyond the mRNA vaccines has been highly productive in the ten months it's been in place."

"Tell me how it's different from the mRNA vaccines," the president said.

"For one thing, it's a nasal spray rather than an injectable. Since this is an airborne agent that gets into the body through inhalation, the vaccine works by neutralizing the virus in the nasopharynx before it can get into the bronchial tree and lungs, unlike the previous vaccines. We are hopeful it will prevent transmission to a far larger extent."

"Don't forget, hope is not a strategy, but I do like that it doesn't involve a needle," the president commented. "It's a shame the kids can't vote."

"Just as important," Sulbarry continued, "the animal studies and clinical trials suggest that it will provide more lasting immunity than we've seen with both the Covid-19 and SARS-3 mRNA vaccines. So far, it looks like we're getting at least ten months of robust protection per two doses, and we have reason to hope it may be even longer than that."

"How soon can we get it out?" Chief of Staff Gilbert Stern asked, cutting to the chase.

"We are applying for FDA Emergency Use Authorization as soon as possible. Hopefully within the next couple of weeks."

Knowing that such an authorization would put them one step closer to saving lives — and making a difference to the electorate — the president pushed for the critical details. "How fast can we make it and get it up people's noses?"

"Not as fast as we'd like, I'm afraid. Up until now, we've had an approved live attenuated mucosal vaccine for flu only, never one for a coronavirus. The new vaccine is a spray squirted into the nose. Manufacturing at scale is going to be a challenge. Several Covid mucosal vaccines have been developed outside of the country, but we don't have confidence they are better than the current Covid vaccine. Had we invested in this kind of research and development earlier, based on what we learned from Covid, we'd be in a better position today. But we can't change the past, so we are forging ahead as best we can."

"For more than a year I've been hearing about all the things we should have learned from Covid, and all the things we should have done but didn't, and haven't." The president sighed. "I know: political priorities. Everyone *was tired of hearing about pandemics. I wonder if they'll be saying the same things about what* we *didn't learn or didn't do in this pandemic — with far greater consequences, I might add." He turned to Stern. "Gil, I understand we're getting pushback even before this new vaccine is ready."*

"Our polling does show that when people are asked if they would take a new vaccine that was more effective than the current one, with more lasting protection and less risk of transmission, a significant percentage — less than half but still significant — said they would be skeptical, based on initial claims made for previous coronavirus vaccines. Others expressed concern that a new type of vaccine might not be safe."

"It is definitely a lot safer than taking your chances with SARS-3," Sulbarry noted. "They should look at how many people this virus is still killing every week!"

"Unfortunately, perception is a lot more powerful than reality, as we all know," Stern responded.

"Okay, look, we've got to get past that," the president said. "This may be the most important thing we do in this administration."

"I agree," said Sulbarry. "And your strong leadership is crucial."

"Deann, how do we show this leadership? What say you?"

Up to this point, press secretary Morgan had been uncharacteristically quiet. "Well, sir," she began, "I do have some ideas..."

From her seat at the command station in the middle of the emergency department, Erin Thomas glanced up at the census board, listing all the patients being seen. It was completely full, with extra names and status notes crammed in. Almost all were SARS-3 patients, with a smattering of the accident injuries, cases of chest pain, and pregnant women in labor they used to see. And before the pandemic, a quarter or more of the patients seen in her department were there for acute mental health issues. Their absence now was of little consolation to Thomas and her team. They knew the need for mental health care hadn't decreased; if anything, it was likely much higher than before the pandemic. But now these patients were simply falling through the cracks of the healthcare system.

Overall emergency department staffing was down so severely that Thomas had had to bring in people who were still sick, recovering from SARS-3 themselves but able and willing to work, wearing N95 respirators to protect their patients from possible transmission if they were still infectious. Of course, this itself was an issue, as the hospital was facing shortages of N95s along with many other supplies, ranging from saline to critical drugs.

From what Thomas was hearing from her colleagues, some other hospitals had it even worse, having to institute combat-like triage, concentrating on the patients with the best chances of survival and giving only the most basic palliative care to those they could tell weren't going to make it. She'd seen on the news that in other parts of the country, hospitals were telling ambulance personnel not to transport—or even give precious oxygen to—patients unlikely to survive. It seemed they were losing the war with the virus. She wondered how long this nightmare was going to last.

As bad as conditions were, she worried about the patients they weren't seeing, in addition to those with acute mental health issues: the heart attacks, strokes, and head traumas afraid to come in for fear of catching the virus. She'd heard the same fears from other departments in the hospital. Cancer patients weren't being seen because of the shortage of doctors, nurses, and techs, as well as the drugs that were on back order that they couldn't get. All but the most critical emergency surgeries had been put on hold, and the community drug and alcohol rehabilitation programs had all but stopped functioning. She'd seen reports online of people holding up drugstores and pharmacies at gunpoint, the way people used to hold up banks. But they weren't looking for painkillers only. Some were desperate for standard lifesaving medications for their loved ones, ranging from insulin to immuno-suppressive drugs for those who'd had an organ transplant. Often, the targeted stores didn't even have what they needed, thanks to the breakdown of the global supply chain.

And she worried about the standard of care they were providing. Not only were staff members now chronically pushed past their limits, working far too many hours and experiencing serious exhaustion and burnout; they'd needed traveling nurses to supplement personnel, very few of whom had actual emergency medicine experience. On-the-job, real-time training was both time-consuming and terrifying given the potential consequences of even a single mistake.

Too worried to get the good night's sleep she so badly needed, she finally understood the meaning of the term "bone-tired." She also had to admit to herself that, for the first time in her professional life, she was afraid.

At the beginning of the pandemic, before anyone knew what was brewing, by the time Curt Ashworth's parents had driven him halfway back to their home in Albany, Georgia, after his volunteering stint in Kenya, he was laboring to breathe. Rather than bring their son home, Caleb and Helen Ashworth had taken him straight to the emergency room. He'd spent the next nine days in the ICU, and another week in the medical ward, before he was stable enough to go home, too weak still to attend the next college semester. As it turned out, he didn't miss much, as midway through the semester most colleges around the country stopped holding in-person classes, and so many professors and teaching fellows became ill that even virtual learning was often impossible. Now, living in a mostly empty dorm, enrolled in classes that only sometimes were able to meet, Curt had seen the toll SARS-3 had taken around the world. As one of the earliest cases, he knew how lucky he was to be alive. He still had nightmares about suffocating in the ICU, and they didn't end when he awoke, as he thought of all the people he might have unwittingly infected during his journey home and how many of them may have died.

He'd called home last weekend to coordinate plans for Thanksgiving break and learned that his mother hadn't been feeling well. Though he was concerned, she reassured him she didn't think it was SARS-3. They'd all had the two-dose mRNA vaccine by now, and she wasn't having any trouble breathing. As a military spouse, she was used to toughing it out, especially when her husband was away on deployment and she had to manage the household by herself. Both Curt and his father pushed their concerns aside.

But when Caleb got home from work late Monday afternoon, he found Helen curled up on the living room couch, shivering under several blankets, unresponsive until he shook her to wake her up. Despite the shivering, she felt hot to the touch, and when he took her temperature, the thermometer read 104. She could hardly support herself as he practically carried her to the car and rushed to the hospital.

When they reached the ER, the triage nurse realized Helen was in critical condition and brought her to a treatment room right away. Vital signs and blood tests led to a quick diagnosis of sepsis, a life-threatening condition where bacteria were growing in her blood. Unfortunately, the hospital had gone through its supply of saline bags, so the staff couldn't immediately administer IV antibiotics as they typically would have. They put out an emergency call to other area facilities, but everyone had been scrambling for the same supplies for weeks. By the time appropriate treatment could be started, the sepsis had already progressed too far. Helen Ashworth passed away the next afternoon, never having left the emergency department.

It would not be recorded as a SARS-3 statistic, but there was no doubt among the physicians and nurses about what had caused her death.

———————

The meeting in WHO director-general Kolawole Adebayo's office in Geneva had been tense from the outset. He had called in Jeremy Davies of the Health Emergencies Programme when he got the projections on global manufacturing capacity for the mucosal vaccine that the Americans, British, and Germans had developed. Those three nations and other high-income countries would probably be able to secure enough to vaccinate most of their populations; the low- and middle-income countries were another matter.

"If the Pandemic Preparedness Treaty and the Patent Waiver Agreement were both put in place after Covid, shouldn't there be an immediate sharing of the technology?" Adebayo asked.

"In theory, yes," Davies replied. "But none of that matters if the places we want to manufacture the vaccine don't have the capability to produce it in-country. Kola, we've been trying to establish multiuse manufacturing operations in Asia and Africa, to make flu vaccine and some polio vaccine in non-pandemic times, but you know how that's gone. And we can't just start building factories from scratch in the midst of this thing."

Adebayo shook his head. "You would think the Covid disaster would have been a wake-up call for the world, but . . . " He let his voice trail off.

"Even if we had that factory capacity," Davies went on, "it's useless if those countries don't have the expertise to staff the factories and then distribute the vaccine. Gavi has been more than willing to put serious resources into this, but where do you start at this point?"

"It's a question of where we *should have* started," *the director-general said sadly.*

"From NBC News in Washington, the longest-running show in television history, this is Meet the Press, *with Jonathan Goodwin."*

"Thanks for spending part of your Sunday with us. Many of you will remember that the president made a public event out of being vaccinated with the SARS-3 mRNA vaccine when it became available. This past Tuesday, in a televised event at the Oval Office, amid skepticism that the new nasal spray vaccine was both safe and effective, the president became the first American to take it, just a day after it received Emergency Use Authorization from the Food and Drug Administration. In a leadership coup that some condemned as a political stunt but most commentators hailed, the president invited the four living former presidents to join him in receiving the new spray vaccine. All four accepted. The president said he hoped every American would soon have the same opportunity for protection from SARS-3 and that they would avail themselves of that opportunity without fear or political influence.

"In other news this week . . . "

DUE TO ITS MAGNITUDE in the annals of public health, we have made frequent mention of the 1918 influenza pandemic, by numbers the most lethal and destructive pandemic in recorded history. Where you wouldn't have heard much mention, if you were alive at the time, was from the White House. President Woodrow Wilson, concerned that focus on the flu pandemic would interfere with recruitment, morale, and pursuit of the war effort, essentially publicly ignored it. This has its own irony, since historians have come to believe, from considerable evidence, that during the 1919 Versailles peace conference, Wilson was suffering from a serious case of the H1N1 influenza that affected not only his stamina but also his mental faculties and judgment, causing him to accede to the ruinous terms of the treaty that ultimately contributed to the conditions in Germany that led to the Second World War.[1]

Though there were no effective pharmaceutical interventions for flu that early in the twentieth century, when we didn't know it was a virus, had Wilson demonstrated proper leadership in what turned out to be a far more devastating killer than the war that provided its

backdrop, cities and states likely would have taken more robust measures to alert their populations to the danger and physically prevent
transmission.

There is no substitute for political leadership and courage
during a pandemic. In 2020, President Donald Trump repeatedly
minimized the dangers and impact of SARS-CoV-2 in his public
statements while conceding privately to journalist Bob Woodward in
a February 7 phone call that "this is deadly stuff . . . [2] You just breathe
the air. That's how it's passed," and acknowledging, "It's also more
deadly than . . . even your strenuous flu."[3]

Confronted months later with Woodward's revelation, Trump
rationalized that he didn't want to scare people.[4] Even if we accept a
positive motive for the contradiction, this is not good or effective
political leadership. As we outlined in Chapter Five, history has
shown time and again that a population under threat reacts more
effectively and positively if they are given the truth, even if the reality
is terrifying.

When, days after becoming prime minister in May 1940, Winston Churchill in the House of Commons declared that he had nothing to offer but blood, toil, tears, and sweat, he was preparing the
British nation for the ordeal he knew was ahead. At the same time,
though, he confirmed the ultimate goal of "Victory at all costs—
Victory in spite of all terror—Victory, however long and hard the
road may be," affirmed the existential nature of the struggle, "for
without victory, there is no survival. Let that be realized; no survival
for the British Empire," and, most importantly, ended by rallying the
population with a note of optimism that with cooperation, they
would eventually get through it: "I feel sure that our cause will not be
suffered to fail among men. At this time I feel entitled to claim the
aid of all, and I say, 'come then, let us go forward together with our
united strength.'"[5]

As famed CBS correspondent Edward R. Murrow said of Churchill, "He mobilized the English language and sent it into battle to
steady his fellow countrymen and hearten those Europeans upon
whom the long dark night of tyranny had descended."[6] In times of

crisis, the right words from the right leader can have enormous impact and literally save lives.

Trump's words and actions during Covid, by contrast, were contradictory and confusing. While downplaying the seriousness of the pandemic, he declared a state of emergency on March 13, 2020,[7] and urged people to stay at home. He agreed to an enormous aid package, the Coronavirus Aid, Relief, and Economic Security (CARES) Act, a $2.2 trillion economic stimulus bill,[8] that gave out money with little regard to actual need. His White House spearheaded the $26 billion Operation Warp Speed[9] to bring the NIH, Department of Defense, and the private pharmaceutical industry together to rapidly develop a SARS-CoV-2 vaccine.[10] Trump brought in Stanford University radiologist Dr. Scott Atlas because he liked Atlas's laissez-faire attitude and assertions that the death projections were being overplayed and that herd immunity would take care of the problem without much economic disruption.[11] All of Atlas's theories turned out to be wrong. With the backing of some Fox News personalities, Trump pushed the value of the antimalarial drug hydroxychloroquine, which ultimately proved useless against SARS-CoV-2.[12] In April, he went so far as to suggest that ultraviolet light applied inside the body or injections of bleach might effectively combat the virus.[13] At the same time, Trump curtailed communications from actual experts in government.[14]

His own messaging, verbal and nonverbal, was damaging to public confidence and the seriousness with which many took the public health crisis. Upon coming down with a serious case of SARS-CoV-2 himself in the beginning of October 2020, following an unprotected White House gathering celebrating the nomination of Amy Coney Barrett to the Supreme Court, Trump was flown by helicopter to Walter Reed National Military Medical Center in Bethesda, Maryland, where he received state-of-the-art care that probably saved his life. Back at the White House, in an image that has become part of our national memory, the president came out on the Truman Balcony and defiantly ripped off his mask. The symbolism was lost on no one.

One of the pandemic's great inflection points was when Trump and his close advisers decided and announced that the federal government would not hold primary responsibility for management of the pandemic and critical medical supply and respiratory protection device resource allocation and distribution. Rather, it would be up to the individual states, and their governors, to lead the effort. This had an immediate, chaotic effect, forcing the governors to bid against one another for critical supplies and equipment, raising their cost for everyone. What would have made much more sense and created a greater level of national cohesion and confidence was the organization of a national purchasing and distribution system based on the real-time needs of each region.

As stated in *Lessons from the Covid War: An Investigative Report,* a 2023 book overseen and edited by Philip Zelikow of the University of Virginia and produced by the Covid Crisis Group, of which I was a member, "From the top, with President Trump, the administration had no real will to offer federal executive direction to the field, to offset the weaknesses of the outdated American public health structure. The administration abdicated its wartime responsibility to lead. It left the battlefield, and the war strategy, to state militias (led by their governors) and ad hocism at the local level."[15]

From the moment he was elected president, Joe Biden demonstrated a far more consistent approach, taking the pandemic seriously and showing real empathy for its sufferers. I saw this firsthand, as I was appointed to the Biden-Harris Transition Covid-19 Advisory Board. Unlike the previous administration, the Biden administration supported a science-based public health response.

As soon as he assumed office, Biden proposed the American Rescue Plan, a $1.9 trillion stimulus package that was passed by Congress in March 2021.[16] It provided numerous initiatives to support vulnerable populations inordinately affected by the pandemic and established a robust system for producing and distributing vaccines, tests, and critical supplies. Among other aims, the plan was particularly concerned with getting all schools reopened as safely as possible; improving indoor air quality; helping states and localities increase healthcare staffing and expertise; identifying new viral strains and coming up with effective antiviral medical countermeasures; providing emergency paid and medical leave while reimbursing employers for the costs of that leave; and extending unemployment and health insurance coverage affected by the pandemic. The plan also addressed homelessness and food insecurity, especially for children.

Was the plan perfect? Of course not. As with the CARES Act, it was a blunt instrument attempting to deal with a multitude of problems quickly. But it did show an acceptance that the personal and economic effects of Covid were real and widespread, and I think it demonstrated a level of compassion and realism characteristic of that president and his administration.

On the other side of the ledger, however, on September 18, 2022, during an interview on the long-running CBS program *60 Minutes,* President Biden declared, "The pandemic is over. We still have a problem with Covid. We're still doing a lot of work on it. It's—but the pandemic is over."[17] While the major impact of the pandemic was behind us in September 2022, an additional 114,000 Americans would die from Covid by the end of 2023.[18] As much as Biden wanted with the *60 Minutes* interview to reassure the public and show that his administration's policies had worked, the statement caused many who had not yet let down their guard to believe they could go back to normal. That attitude was easy to understand; by then, we all suffered from pandemic fatigue.

At the same time, the CDC continued to provide confusing and contradictory recommendations to the public about effective respiratory protection, the need for N95s, and how many days after testing positive one could go back to work—to say nothing of its strong support of contact tracing, before reversing itself within months. And despite the billions of dollars that went to schools, little was spent on improving air ventilation or even research on what it would take to make those indoor environments safer. At the end of 2023, the NIH had spent $1.15 billion on studying long Covid, with virtually nothing to show on understanding its causes or how to treat it.[19]

We've noted that the Covid pandemic is a major stress test for modern society. It exposed domestic and global vulnerabilities to the economy, healthcare systems, and international supply chains for critical materials and products, including drugs and medical supplies. It revealed the alarming underlying health condition of Americans, a collective vulnerability that led to one of the highest Covid mortality rates of any high-income country. And there has been no 9/11-type government-authorized commission established to catalog and understand what was done right and wrong and what lessons we can take away from the experience in planning for future pandemics, including the Big One, which will be the stress test for the world.

THE POLITICS OF THE PANDEMIC

I've been the object of a fair amount of criticism in my public health career, but until the Covid-19 pandemic, I'd never received a death threat. During that pandemic, I received many. And I'm not alone. Almost every prominent public health official I know has had threats made on his or her life simply for stating the best scientific knowledge, information, and insight we each had at the time. I routinely turned over such threats to our local law enforcement, which investigated them and, when necessary, passed them along to federal authorities.

Now, even if you disagree with one of us (and we often disagreed with one another), what kind of society are we living in where some individuals feel it is okay to threaten to kill a scientist, doctor, or public official for expressing an opinion, or even a recommendation, aimed at the public good? What do they think is motivating us, some megalomaniacal attempt to control or mislead the public for our own purposes? And what purpose could that possibly be? We ask these questions not because we are unaware of the conspiracy theorists who post ready answers on social media, but because the situation we find ourselves in violates everything we work for.

At a time when we desperately needed to be focusing on how to combat the SARS-CoV-2 virus itself, these same conspiracy theorists and their fellow travelers were obsessed with proving the virus originated in the lab in Wuhan from some sinister gain-of-function experiment or an accidental release from the lab rather than by a spillover from infected animals to humans at the local wet markets. Gain-of-function research involves genetic enhancement of a virus to predict and prepare for dangerous evolutionary changes. Make no mistake, both a lab leak and a spillover event must be taken seriously. A pandemic begins when the first human case is infected from an exposure to a novel influenza or coronavirus, or a novel virus that has similar transmission characteristics to these two. The virus could not "care" less which way it gets into the lungs of the first human it infects. So, we must consider both sources of a potential pandemic virus and be prepared to deal with them.

I served on the National Science Advisory Board for Biosecurity from 2005 to 2014. It is a panel of experts that reports to the HHS secretary, advising the US government on how to prevent research in the life sciences from causing harm, including research that might accidentally release an infectious agent to humans or animals. During my tenure on the advisory board I raised concerns about lab safety and experiments using selected influenza viruses in gain-of-function studies. No one could call me soft on the kind of research being conducted in 2019 in Wuhan. With this background information, I hope my conclusions about the source of SARS-CoV-2 carry some credibility.

I have carefully reviewed every available study, report, or media story regarding the debate about the source of the virus. I'm convinced that we will never know the source. While I think the collective information that we have provides more support for a spillover being responsible for its introduction into humans, there is also limited evidence to support a possible lab leak. But at this point the issue is more cultural than scientific and has become a major distraction from almost every effort throughout the world to assess what lessons we should have learned from our Covid experience to prepare for the Big One. In short, it's time to stop the bitter debate about the source of this horrible pandemic and get on with preparing for both future lab leaks *and* spillovers of influenza or coronaviruses, or any other as yet unknown viral agent capable of causing a pandemic.

Despite what we've said about the erosion of trust in evidence-based science coming to the fore during Covid, the phenomenon itself is not new. We have only to go back to *The State of Tennessee v. John Thomas Scopes,* the famous Scopes Trial of 1925, in which a teacher in the small town of Dayton was charged with teaching evolution in his public school classroom, to see it manifested in the American body politic. The state's Butler Act made it illegal to teach evidence-based evolution in state-funded schools. Compared to biblically based views about where humans and other animals came from, the idea of Darwinian evolution was considered such a threat that it had to be banned. This epic test case pitted the agnostic defense lawyer Clarence Darrow against the deeply religious populist politician William Jennings Bryan, who died five days after its conclusion, possibly from the stress of the trial and his own testimony on the witness stand. The jury decided against Scopes, who then lost again on appeal, though the higher court set aside the token fine that had been imposed as his punishment. The repercussions of the Scopes Trial have been debated in the century since, but there can be little doubt its reverberations continue to this day.

There have always been snake oil salesmen looking to profit from the desperation or gullibility of others, but what we are seeing now is different. Anti-science has become a sociopolitical movement of its own. It is as if an entire cohort has forgotten or is ignoring the blessings that science has brought to make our lives better and longer, from antibiotics to vaccines for polio and other diseases, drugs for a wide variety of ailments, and strides in cardiovascular and cancer treatments. The stance, which Peter Hotez, the renowned vaccinologist and pediatrician at Baylor College of Medicine and Texas Children's Hospital, labels anti-science aggression,[20] is unfortunately not limited to the misinformed. It has been promulgated by certain members of Congress, governors, conservative news outlets, and even self-proclaimed experts from academia. Texas, for example, prohibited health facilities funded by the state government from specifically recommending the Covid vaccine.[21] Florida went even further, with its state health department issuing guidance *against* the vaccines.[22]

It is estimated that by the end of 2023, more than 200,000 Americans had died needlessly because they didn't trust the Covid vaccine.[23] And this skepticism has even more problematic downstream effects, such as when the anti-vaccine trend moves into vaccines we've trusted for years, such as those for polio, measles, tetanus, diphtheria, and other diseases. It is beyond sobering to think that in 2000, measles had been eliminated from the United States[24]— a stunning scientific victory—but two decades later, it is back with a vengeance because too many parents have rejected vaccination for this highly infectious and potentially deadly virus.

Interestingly, anti-vaxxers seem to be the one segment of the overall anti-science landscape that has a fair number of young, liberal adherents. Many of the same people who have embraced the environmental movement, concerned for how climate change will affect the world of their children's future, subscribe to the completely discredited view that vaccines cause autism; one popular article even falsely claimed vaccines could render children sterile.

While we can argue in good faith which states and governors fared best during the various phases of Covid, it is a fact that conservative, Republican-dominated states had much higher rates of vaccine-preventable Covid deaths.[25] As Peter Hotez writes in his 2023 book, *The Deadly Rise of Anti-Science: A Scientist's Warning*, "Partisan leanings were strongly associated with the likelihood both to be unvaccinated and to lose one's life to Covid-19. The 'redder' the county in terms of the percentage of Republicans, the higher the loss of life."[26] This is not a political statement; it is rather an objective epidemiological finding.

A 2023 article in the journal *Health Policy OPEN* entitled "The

Politics of COVID-19: Differences between U.S. Red and Blue States in COVID-19 Regulations and Deaths," describing a well-designed study, concludes,

> Diseases have demonstrated no partisan allegiance, past or present. The individual role of citizens is not without consequence, but to ultimately lessen the aversive effects of COVID-19 and other viral threats in the United States, it is necessary to behave collectively. Given the compelling evidence of mass-behavioral mitigation efforts being successful in pandemic remediation, further legislation should focus on best communicating and implementing these strategies across political landscapes. Focusing on effectively implementing mitigation strategies across ideologies should be paramount if communities are to address disease-based threats with minimal loss and aversive outcomes.[27]

Also in 2023, a poll conducted by KFF (formerly the Kaiser Family Foundation) determined that 84 percent of Democrats said they were confident in the safety of Covid vaccines, compared with 36 percent of Republicans.[28] A study published in *JAMA Internal Medicine* the same year found that prior to the Covid vaccine becoming widely available, Republicans and Democrats in Ohio and Florida died at similar rates, but by the end of 2021, Republicans in those states had a 43 percent higher rate of excess deaths than Democrats.[29] Another telling statistic, reported in *JAMA Network Open* in March 2024, captures the association between political leaning and the likelihood that someone would report experiencing an adverse event (AE) from vaccination:

> This cross-sectional study of 620,456 AE reports found that a 10% increase in state Republican voting was associated with a 5% increase in the odds that a COVID-19 vaccine AE would be reported, a 25% increase in the odds that a severe AE would be reported, and a 21% increase in the odds that any reported AE would be severe.[30]

The report concludes:

This cross-sectional study found that the more states were inclined to vote Republican, the more likely their vaccine recipients or their clinicians reported COVID-19 vaccine AEs. These results suggest that either the perception of vaccine AEs or the motivation to report them was associated with political inclination.[31]

The figures come from data in the Vaccine Adverse Event Reporting System (VAERS), a tool used by individuals and clinicians to report AEs, and by those in public health to monitor vaccine safety. Unfortunately, the purpose and limitations of VAERS have been misunderstood by many policymakers and the public. As a result, the potential AEs entered into the system may have nothing to do with the receipt of a vaccine; their occurrence may be coincidental. Consider that each day, millions of Americans experience potentially severe health events such as heart attacks, strokes, seizures, etc. With many thousands vaccinated every day — in particular, young children, for whom it is recommended to receive shots at birth and at two, four, six, twelve, fifteen, and eighteen months of life — it is highly likely that some health event will occur shortly after vaccination that is not related to receipt of the vaccine. For example, we know statistically that a certain number of women will have less than optimal outcomes when they give birth. And a small subset of those will have received a vaccination sometime during their pregnancies. While this is a *correlated* finding, that doesn't mean it is a *causative* finding.

VAERS was designed to allow reporting of possible AEs by both the public and the medical community to allow early detection of AEs not previously recognized to be associated with vaccination. Because potential AEs can be *any* health event occurring around the time of receipt of a vaccine, it's the job of the VAERS staff to vet reports, identify all accounts of similar AEs reported to the system, and determine if there are other comparable reports that require further, immediate consideration. The trouble is, vaccine opponents often cite *every* report made to VAERS as being caused by the vaccine in question. For good or ill, perception is reality.

And "reality" is increasingly becoming a moving target. Against the backdrop of the rapidly developing field of artificial intelligence (AI), as we touched on in Chapter Five, I worry even more going forward about mis- and disinformation campaigns that to the uninformed or already biased will be inseparable from truth. It is already easy for AI to generate convincing and authoritative-sounding—but entirely false—medical journal articles that promote all manner of deadly mischief for ulterior purposes.

In 2024, Dr. Sandro Galea, my distinguished friend and colleague and inaugural Margaret C. Ryan Dean of the planned Washington University School of Public Health, delivered the Gaylord Anderson Memorial Lecture—named after the founder and first dean of the School of Public Health at the University of Minnesota—entitled "Within Reason? Ensuring Public Health Matters in Coming Decades."[32] The lecture was based on his widely acclaimed 2023 book, *Within Reason: A Liberal Public Health for an Illiberal Time*. His statistics and conclusions should be sobering for us all. Tracking "Trust in Pandemic Information Sources, Republicans and Non-Republicans," Sandro shows four charts representing trust in the CDC, NIH, medical experts, and international health agencies from April 2020 through July 2021. In each case, trust among non-Republicans remains fairly consistent, and trust in the NIH and medical experts rises, while trust in all four sources among Republicans falls off markedly.[33] He also shows that trust in public health is generally lower among younger generations than it is among baby boomers.[34]

But Sandro doesn't let the public health establishment off the hook, either. Under "Public Health's Key Shortcomings during the Pandemic," he lists "false certitude, contradiction without acknowledgment, and intolerance of disagreement,"[35] mirroring Peter Sandman's observations in Chapter Five. Sandro explains those shortcomings as due to the complexity of systems, our biases and privileged perspectives, and groupthink and absolutism. I strongly believe his analysis and conclusions are right on the mark. We all have lessons to learn about leadership and policy, and it all ought to be tempered by the humility we stressed earlier.

An August 2023 memo from the Republican Study Committee entitled "Conservative Engagement in Upcoming Debates on Public Health Policy" outlined an advocacy agenda based on Congress's "responsibility to rein in the federal bureaucracy's overreach in public health" and the "Government's Failed Response to COVID." The advocated actions included "rein in the CDC," and "cut the slack" from the NIH.[36]

As someone who has criticized (as well as lauded) both institutions, I understand the need for self-reflection on how we can improve, and sometimes actually overhaul, our efforts in particular, making us more prepared for the Big One. So, if I thought the intent of the Republican Study Commission was actually to make these institutions better and more effective, I'd support their efforts. But unfortunately, I don't have any belief that their advocacy agenda is being pushed in good faith. It seems to be a case of throwing out the proverbial baby with the bathwater. I see it as just one more political assault on public health and an appeal to an anti-science base.

We've noted earlier the initial differences in how the Scandinavian countries reacted to Covid-19, with Sweden essentially staying open and Denmark, Norway, and Finland shutting down to greater or lesser degrees. They all fared about the same in the long run with regard to morbidity and mortality, and better than many other high-income nations, including the United States and the United Kingdom. This was due to one primary factor: social cohesion that led to trust in government and political leaders. All of the Scandinavian countries share a national ethos that everyone is in this together, they all have the same ultimate goal, and there is social benefit in looking out for and cooperating with one's neighbors. Additionally, of note, even though it was one of the most successful countries in dealing with the SARS-CoV-2 virus, Norway, unlike the United States, established a commission to assess what was done, right and wrong, and how they could have done better.[37]

While it is true that Scandinavia differs from the United States in that those countries are far more unified in their cultures and ethnicities than we are, the results, and the lessons, are hard to dispute.

FORECAST MODELING AND MATHEMATICAL MODELING: NOT KNOWING THE DIFFERENCE IS A REAL PROBLEM

From the earliest days of the Covid pandemic, one of the most frequently asked questions was, "When will it end?" Modeling tools varied greatly in their ability to give us reliable and timely predictions of what the future course of the Covid pandemic would look like. But "crystal ball" estimates based on complicated statistical models that predicted many months into the future were essentially useless. In large part that was due to a fundamental misunderstanding of what statistical, mathematical modeling is all about in predicting future events. Yet throughout the pandemic, the media couldn't get enough of modeling groups telling the public what they could expect the pandemic to look like months in advance.

What went wrong? Almost no one, from policymakers, statisticians, epidemiologists, and modelers to the media and surely the public, understood the difference between short-range forecast models and mathematical models. What is the difference? Our earlier analogy of predicting weather and climate is relevant here.

Weather forecast modeling focuses on what will happen in the coming days and for the next several weeks. The starting point is an existing expansive network of reliable observations. These are the data that national government weather services, academic departments, and private companies have collected over decades from a local level all the way up to a global level. When all the myriad observations and data points are taken together, it's possible to predict the weather in a given area for a reasonable time frame. To improve the forecasting accuracy, weather scientists have developed what is called ensemble modeling, where a model is run multiple times with the initial conditions altered slightly, resulting in a forecast with the most likely probability. It also refers to adding different but compatible models to determine a final comprehensive estimate. Infectious disease forecast modeling has been shown to have a similar degree of accuracy. Forecast models could give us fairly reliable predictions of what would happen in the next month with regard to case incidence.

Mathematical modeling, on the other hand, attempts to assess a number of different conditions or factors that might impact a measured outcome such as the number and severity of Covid cases without any, or limited, previous historical data. For example, a mathematical model may consider how the pandemic might look months later based on a range of vaccination estimates, including how many people will get vaccinated and how effective the vaccines might be; the relative protection afforded by different kinds of masking behavior, including the number of people masking and how efficient the types of masks might be; and level of population mixing in public. The model is based on estimates of each of these and other factors, without the modeler necessarily knowing how accurate the estimates are compared to the actual experience. If the input isn't accurate, the output won't be, either.

Unfortunately, in the first eighteen months of the Covid pandemic, the models typically reporting results were mathematical, most of which turned out to be unreliable to the point of being useless, or even harmful.

In public health, as in medicine, we should set as our highest operating principle to do no harm. I strongly believe that without realizing the potential negative impact of their studies, the practitioners of long-range mathematical modeling violated that principle with their early estimates of how the pandemic would unfold in terms of cases, deaths, and the impact of various mitigation efforts. I'm afraid the early predictions of Covid case and death numbers were greatly overestimated and led to many accusing the public health community of exaggerating the risk of Covid and being responsible for forcing policymakers to adopt actions like lockdowns and mask mandates.

I experienced firsthand the blowback from such modeling estimates when researchers at the University of Minnesota School of

Public Health and the Minnesota Department of Health developed mathematical models in early March 2020 to estimate the health impact of Covid in the state. These researchers were not part of CIDRAP, nor did we consult with them. Many assumptions were made as they built their models on how the virus would spread, the number of severely ill cases needing hospitalizations, and how many would die. They also factored in how varying levels of mitigation were employed. Their work was shared with Minnesota governor Tim Walz. When I learned of the modeling study results and the intent of the governor to use them in determining his policy response to the pandemic, I urged the Minnesota Department of Health to pull back. My input carried little weight; the response train had left the station. Governor Walz later told me that hearing such dire numbers was one of the most unsettling and scary moments of his career.

On March 25, 2020, he held a press conference presenting the two scenarios provided by the modeling group.[38] One projection concluded that Covid cases in Minnesota would peak around April 26 if no mitigating steps were taken, with the number of deaths reaching as high as 74,000.[39] The other scenario determined that with mitigation measures like a major lockdown, the number of deaths would peak around June 29, three months later, at 50,000 to 55,000.[40]

So, what did happen? By *November 2024*, more than *four years* later, in-state reported deaths totaled 12,800.[41] While this is certainly a tragic outcome, it is nowhere near the extreme predictions of either scenario, and it surely didn't enhance the credibility of the public health establishment.

The grossly inaccurate death estimates gave "proof" to every public health critic that we couldn't be trusted, leaving the governor high and dry as he depended on academics from his state's flagship university. Personally, I feel the pain of this experience today. As a professor at the School of Public Health, I am frequently identified as one of those who put out the "wild estimates" that resulted in state actions that sank the Minnesota economy.

This experience and others like it in the United States and Europe regarding eye-popping, but very preliminary, modeling results should serve as a learning experience for the release of such data for highly charged public policy and health issues. Had my Minnesota colleagues vetted their initial work product with infectious disease experts in outbreak response before releasing it, many of the shortcomings in data analysis could have been avoided with these models. And to public policy and elected officials: Never make major decisions on emergency public health action with input from a single modeling group, and always determine if the model is a forecast based on a large volume of preexisting data; I don't care who it is. As in medicine, it's a good idea to get a second opinion.

With infectious disease mathematical modeling, it is all too easy to get out over your skis. In other words, don't pretend you can predict some future event or outcome when major impacting factors can and likely will change the course of the pandemic in the near future. For example, several prominent modeling groups made predictions during Covid of the number of cases and deaths up to six months out. Imagine estimating with any accuracy how much rain will fall in Minneapolis on a given day six months from now. With Covid, the major wild card turned out to be the emergence of new SARS-CoV-2 variants. Though the first three of eight peaks of cases observed in the pandemic—which occurred in 2020 and early 2021—were not associated with new variants, the last five clearly were. While vaccination levels are surely important to the number of serious illnesses, hospitalizations, and deaths, the primary factor impacting the course of the pandemic was variant emergence. How could any modeler anticipate the next variant to show up in six months and change the course of the pandemic, one way or the other? One of the lessons to learn from the natural history of the Covid pandemic is that, like weather forecasting, any modeling prediction more than thirty days out is highly vulnerable to changes and can't be reliably used.

One footnote: When I predicted in April 2020 that the United States could experience 800,000 deaths in the next eighteen months, and we ended up being painfully close to that number, unlike the modelers, I didn't use a black box analytic approach, making a series of statistical assumptions. On paper, I estimated the number of Minnesota residents who would get infected in the period and considered different mortality rates. Really very simple, but it was more accurate than the long-term models that were bandied about in the news.

The instability of long-range mathematical forecasts is exemplified by one of the most cited modeling groups, the Institute for Health Metrics and Evaluation (IHME) at the University of Washington. Its models forecast the numbers of cases and deaths up to five months into the future. In late March 2020, IHME predicted that by August there would be 81,000 deaths;[42] by early April, it revised the estimates down to 60,415;[43] and by late May, the estimate was more than doubled to 143,357.[44] On July 7, 2020, a press release predicted 208,255 deaths (range: 186,087 to 244,541) in the United States by November 1, 2020.[45] This number was eclipsed by early October—nearly a month earlier than forecast.[46]

Another modeling lesson learned is that the media tends to favor the same groups for model predictions without later reconciling what they forecast with what actually happened. This would be easy to do by merely going back to previous print, radio, or TV interviews and asking, "What did you predict six months ago, and how is it playing out now?" I never saw that done.

And yet another lesson is that model limitations should be clearly communicated to stakeholders. For example, the July 7, 2020, IHME model, which received major national news coverage, predicted that if at least 95 percent of people wore masks in public, deaths would drop by approximately 20 percent to 162,808 (range: 157,217 to 171,193) in the United States.[47] Similarly, in October 2020, the IHME model projected that 130,000 deaths could be averted by February 28, 2021, in a setting of universal mask use. However, the modelers failed to factor in the type of mask or cloth face covering. The IHME model relied on synthesizing data from different published studies (meta-analyses), despite the fact that these studies assessed masks of various types—and therefore various efficacies—

against numerous respiratory viruses.[48] Combining the results of studies on respirators, surgical masks, and cloth face coverings against a range of pathogens didn't give the modelers or the public an accurate sense of the effectiveness of each, and these results certainly could not produce an accurate estimate on the effectiveness of universal mask use. This important nuance was not conveyed by the media.

Better approaches eventually emerged from the Covid pandemic modeling experience. In the early days of the pandemic, the CDC, like many other agencies, organizations, and academic centers, put out the unreliable mathematical modeling estimates. But with time, CDC modelers came to understand the weaknesses of these types of models. By 2021, they began using data from pandemic epidemiologic investigations and estimates from more than twenty other forecast models developed by researchers around the world to put into their own forecast models. And they stopped making estimates more than thirty days out. The thirty-day models proved highly reliable.

This newly established modeling expertise and approach were made part of the CDC's Center for Forecasting and Outbreak Analytics, with initial funding of $200 million from the American Rescue Plan, in August 2021.[49] It has been instrumental in supporting the modernization of infectious disease forecast modeling.

ECONOMIC POLICY

After Covid, we can't say we don't understand the cost of not being prepared, or what a pandemic can wreak in terms of morbidity and mortality, or its effect on the global economy. Clearly, it is worth billions of dollars now to save trillions later—not to mention hundreds of millions of lives—down the road. And our plans must be based on both scientific and political reality. "A + B + Miracle = Solution" is not a realistic formula for success.

We do not claim to be economic experts, but when planning for the Big One, based on the experience of Covid-19, certain important themes emerge. The first point to acknowledge is that a pandemic like Covid-19, or others to come, has dramatically disparate and unequal effects on different sectors of the population.

As Sandro showed in his lecture, divided by income quartiles, in the top quarter, 62 percent were able to work remotely during Covid. In the second-highest quarter, that number dropped to 37 percent. In the third, it was down to 20 percent. And for the lowest 25 percent of earners, only 9 percent were able to work remotely.[50] This has a significant effect on health and safety, not only for the workers themselves but also for their families and those they live with, especially in the time before vaccines and other effective medical countermeasures become available.

In the United States in 2020, the occupation groups with the highest number of deaths from the virus were, in this order: transportation and material moving, construction and extraction, and production.[51] None of these occupations could be performed from the safety of the worker's home. These high-risk essential jobs were often among the lowest paying.

Early in the Covid vaccination program, authorities were criticized for not making low-income and minority populations priorities for vaccination, as they were experiencing the most serious effects from the virus and many were working essential jobs that couldn't be done from home. Yet, as we have noted, if they had been prioritized to a greater extent, we can imagine that the decision-makers would have been accused by some of making those groups the test subjects for the more affluent population. This underscores how difficult it is to remove politics from pandemic policy, and how critical it is to have effective and responsible communication from spokespeople trusted by each community. That fact doesn't, however, relieve officials and public health authorities from making the hard decisions.

As a matter of sound policy, the people who perform the jobs essential to the functioning of society need to be prioritized as we prepare for future pandemics. They should be in the first cohort to receive personal protective equipment, vaccines, and other medical countermeasures, subsidies for childcare, supplemental health insurance, extended sick leave, and hazard pay commensurate with the danger of their jobs. If they have to be out for sickness caused by the pandemic, they should be guaranteed their jobs when they are well enough to work again. It is in all of our interests that these essential workers be treated fairly so that they continue to go to work when the greater good requires it.

And then there is the economy as a whole.

Neel Kashkari is president of the Federal Reserve Bank of Minneapolis and someone I have come to rely on for his knowledge and insight. He served in the administrations of both presidents George W. Bush and Barack Obama, and his mandate as well as his personal integrity make him nonpolitical in his policies and decision-making, as all Fed officers are expected to be. He observed, "Even the countries you and other health experts identified as having managed Covid well still suffered economically. The virus was still in charge and driving all economic challenges. You actually had thoughtful policymakers doing the right things early on, and it still overwhelmed them. The only good news through all of this is we got good vaccines earlier than we anticipated, so we could open up the economy earlier than expected. But some of the stimuli we set up ended up being too much, so you had reopening, disrupted supply chains and lots of spending, driving inflation, though too big a stimulus is understandable at the time. We learned from [the financial crisis of] 2008, when we had undersized everything."[52]

This idea of the massive spending, more than $4.1 trillion with the CARES Act and American Rescue Plan, leading to higher rates of inflation, comes up repeatedly in conversations with economists, business leaders, and other students of fiscal policy. Jason Furman is Aetna Professor of the Practice of Economic Policy at Harvard's Kennedy School of Government and the Department of Economics at Harvard University. He agrees with Neel about the good intentions of the policymakers, with similar caveats. "Broadly speaking, the United States and other advanced economies did pretty well in their economic responses.... Back in the end of February and beginning of March 2020, it looked like there might be a major financial crisis. They completely stopped that from happening, and the recovery was relatively rapid when Covid passed.... One respect in which I think it would be good to improve upon the response is that we threw tons of money at it, and I don't think we can afford to throw $5 trillion every time something like this happens. And so, getting better at targeting would reduce some of the excesses of spending that I think did have bad side effects in terms of contributing to the inflation we saw after."[53]

What was the problem? Simply stated, the government—in both the first Trump and the Biden administrations and the Congress—was trying to create responsive programs in virtually real time. This, obviously, is the opposite of preparing and planning ahead.

Jason continues, "I think some of the issues, totally understandably, were that in the middle of a pandemic, they couldn't invent new things. And so, it was a lot easier to send money to everyone than to send money to some people. It was a lot easier to set up a PPP [Paycheck Protection Program] with almost no eligibility criteria than to figure out how to build and do the eligibility criteria.... Building some of these systems in advance, I think, will let us target our interventions better next time and get most of the good we had here, but at a lower cost and with fewer of the side effects."[54]

One idea is business continuity insurance, outlined in a 2021 Aspen Institute Economic Strategy Group report entitled "Business Continuity Insurance in the Next Disaster," authored by economists Samuel Hanson and Adi Sunderam of Harvard, and Eric Zwick of the University of Chicago. They explain,

> The design of this proposed policy takes seriously the challenge of targeting business support toward firms with the highest private benefit and social insurance value relative to program cost. The program targets assistance to firms whose operations are severely affected by a current shock, that are unable to smooth the shock on their own, and for which bankruptcies would create substantial spillovers.[55]

Among other targeted interventions that economists we spoke to advocate are automatic stabilizers in unemployment systems that kick in when income for those on the margins falls below a specified level or the unemployment rate reaches a certain threshold. The system should also be tied to the level of harm that the pandemic is doing to the economy and society. We can easily imagine that if Covid had the approximately 30 percent fatality rate of MERS, the response would have differed substantially.

"The goal should have been to protect people," Jason says. "To protect their lives and protect their ability to have their living standards. I think in 2020 it was fine, but it went further in 2021, and I think it became more of a problem because then there was more need of people to return to jobs. And one thing you see, the United States had a GDP recovery that was faster than peer countries, but it had an employment recovery that was slower."[56]

"Europe had its own set of solutions," Neel points out. "Some countries have a subsidy where the government will pay an employer to keep you employed, so they didn't lay as many off."[57] It also could prevent widespread collapse of small businesses.

That seems to be a model the United States could plan to implement in future pandemics. Jason explains, "In most European countries, they said, 'You can't work because of Covid? Your employer will keep paying you. You will continue to be employed. And the government will send a check covering the cost of your salary to the employer.' And so, you're a restaurant in London, you kept the people on, you kept paying them, and then on the back end, the restaurant got reimbursed by the government. In the United States, the employer laid you off. Now you're unemployed and you get a check directly from the government. Those accomplish somewhat similar things [in terms] of providing the support for people. But it looks like the European system was better at keeping people connected to their employers and jobs. They had some of the shortages—labor shortages—that we had in the United States in 2021, but they had less of them. And they had more people that were able to be reactivated more quickly."[58]

This is not only economic policy but social policy as well, Jason states. "The pandemic sort of ripped apart the fabric of a set of employment relationships and a set of business relationships with each other, and you want to preserve as many of those as possible. That doesn't mean you want people going to work in restaurants [at the height of viral transmission]. There were very good reasons to have that not happen in 2020. But you wanted to be able to reopen quickly, as soon as it was safe to do so, or as soon as it's an acceptable risk to do so. We did a decent job of that, but I think Europe did a better job of that on the employment side."[59]

What should be done, Jason suggests, is to have in place the kinds of employment retention systems that worked well in places like Germany and Britain. While this plan would cost considerably less than the roughly trillion dollars that PPP ended up costing, it would still be expensive and increase government debt. But if it were planned for ahead of the next pandemic, it could be supported by having businesses pay a small insurance premium, similar to member banks paying into the Federal Deposit Insurance Corporation.

The need for planning extends outside the United States, as Covid demonstrated. While, despite often severe impediments to the global supply chain, high-income nations were able to lessen the pandemic's economic impact on their populations, even at the cost of resulting inflation, for the low- and even some middle-income countries, it was necessary for international organizations to step in. The World Bank's website stated,

> COVID-19 has dealt a major blow to the world's poorest countries, causing a recession in 2020 that is estimated to have pushed more than 100 million people into extreme poverty.
>
> At the start of the pandemic, the World Bank and the International Monetary Fund urged the G20 to set up the DSSI [Debt Service Suspension Initiative]. Established in May 2020, the DSSI helped countries concentrate their resources on fighting the pandemic and safeguarding the lives and livelihoods of millions of the most vulnerable people. Forty-eight out of 73 eligible countries participated in the initiative before it expired at the end of December 2021. From

May 2020 to December 2021, the initiative suspended $12.9 billion in debt-service payments owed by participating countries to their creditors, according to the latest estimates.[60]

In return, the affected countries committed to increase domestic spending on social, health, and economic enterprises to respond to the crisis. Again, this type of international effort will be crucial when the Big One hits, so it should be planned for.

A BROKEN HEALTHCARE SYSTEM

Simply stated, our healthcare system is broken. While the United States leads the world in medical research and pharmaceutical development, the actual care and treatment afforded to many millions of Americans, and not simply the poor or those without adequate health insurance, is woefully inadequate compared to numerous other high-income countries.

We have a massive healthcare system in this country, in terms of financial investment, healthcare delivery infrastructure, and employed professionals. In 2022, the United States had one of the highest costs of healthcare in the world, with spending reaching $4.5 trillion.[61] Meanwhile, the average cost of healthcare per person in other high-income countries was half as much as ours.[62] And our spending continues to rise relative to the size of our economy. It has increased over the past fifty years from 7 percent of GDP[63] to 17.3 percent and is estimated to reach 19.7 percent by 2032,[64] due to our aging population.

We can argue about the relative merits of our multifaceted and disparate system versus single-payer arrangements like Canada's or Britain's National Health, but according to the Commonwealth Fund, a foundation whose mission is to promote better and more equitable healthcare, the United States receives the least return on its investment compared with the ten other high-income nations in the fund's study.[65] This inefficiency, if we want to call it that, has profound implications in preparing for and facing the Big One.

Before the Covid pandemic, hospitals across the country were eliminating beds, particularly pediatric, for cost saving and were already facing workforce shortages of doctors, nurses, and support personnel.[66] The workforce shortage was amplified by the estimated 3,607 healthcare workers who died from Covid in the first year of the pandemic.[67] The conditions Dr. Erin Thomas and her team struggle with in their emergency department in our scenario illustrate how the additional challenges of a pandemic push resources—including personnel—to the brink and beyond. While some healthcare workers died from SARS-CoV-2, and many were temporarily incapacitated, among the greatest issues was burnout. They were simply overburdened, physically, emotionally, and numerically, and the situation did not fully recover as Covid eventually waned. Indeed, it remains a problem today.

One of the often-criticized aspects of the American system is the emphasis on fee-for-service rather than "value-based care" that focuses on efficiency and overall outcomes. This latter model relies more on primary care and therefore gets people into the system earlier in a disease process. It also attempts to get people without adequate means or insurance routine care and preventive treatment before their conditions deteriorate or become more complex and expensive to deal with.

Dr. Brad Spellberg is chief medical officer at Los Angeles General Medical Center, one of the largest public hospitals and medical training facilities in the United States. Brad is an internationally recognized authority on antibiotic resistance and stewardship, demonstrated in his book *Rising Plague*. He is also a hands-on expert on what ails the American healthcare system, elaborated in his 2020 book, *Broken, Bankrupt, and Dying*. He told us, "The fee-for-service system is a failure as a model of healthcare delivery. We're [Los Angeles General Medical Center] a government-run health system. You know what our hospital's net operating margin was last year? It was *minus twenty-four percent*! Because we're not here to make money. The entire ethos and principle of this health system that I work in is to provide the most care to the most people possible with the fixed resources available. So, maybe one of the fundamental things we've learned is: Why the hell is the rest of the US healthcare system still based on a fee-for-service model?"[68]

One concept receiving increased attention is Hospital at Home, or HaH. The idea goes back to 1995 and was conceived by Dr. John Burton of the Johns Hopkins School of Medicine and Dr. Donna Regenstreif of the John A. Hartford Foundation.[69] As a Hopkins bulletin explains, "Hospital at Home allows an older adult with an acute illness to remain in the comfort of their own home while receiving hospital-level care. This model helps older adults avoid common iatrogenic [treatment- or healthcare-environment-related] complications associated with stays in traditional acute care hospitals such as delirium, polypharmacy [adverse effects caused by interaction of multiple drugs], functional decline, and others." Just as significant, HaH results in "cost savings of 19% to 30% compared to traditional inpatient care"[70] and helps alleviate the shortage of hospital beds.

Dr. Eric Topol, director of the Scripps Research Translational Institute, adds that advances in medical technology can add to the HaH equation and further help with hospital overcrowding, especially for intensive care. "Staffing is a vulnerable situation. It doesn't take much to break the whole thing. We are on the brink....When a pandemic hits, we're sunk. If you look at our staffing and beds [during Covid], we fell well below Europe and Asia....You already have sensors that can do ICU-equivalent monitoring in a person's home, so we should prepare for people not having to be in an ICU setting. Medicare gave a waiver [during Covid] so you could get the same reimbursement for treating a person at home. Most health systems said they'd do it, but many didn't. If we wanted to scale for home care, with sensors, etc., to capture data, whereby one doc or nurse could oversee tens of people remotely, and algorithms tell us when people decompensate and need to go to the hospital, we have all the tools to do that remotely at home, but the resources and will aren't there."[71] According to Medicare standards, HaH patients are visited twice a day by medical personnel and monitored remotely.[72]

Brad Spellberg speaks to the transformations that will have to be made in day-to-day institutional healthcare once a pandemic is underway. "You create capacity by redeploying your resources....

You say, 'We're going to refocus all our resources to survive the next twelve months. You're just not going to get elective stuff done. Your colonoscopies aren't going to get done. Your gallbladder isn't going to be removed. Just accept it; we need those resources to be focused on urgent, emergent care....And then we'll reopen as the scenario on the ground dictates that we're able to do. If you're not in a fee-for-service environment, you can actually do that financially very flexibly. Where and when you're in a fee-for-service environment, you are screwed....

"The definition of triage is, how do we take all of the resources available to us and use them in the maximum way to benefit the most people? And [it's easier] if you're practicing in a population-based healthcare system that's already thinking about that."[73]

The difficult but unavoidable conclusion: We cannot adequately prepare for the Big One if we don't revolutionize and rationalize our healthcare system.

UNIVERSAL LESSONS

The lessons here are near universal: The more preparation before the crisis, the better for everyone. That goes for leadership, public trust, social cohesion, and economic benefit. And no matter the economic disparity that exists in the world, we need to keep reminding ourselves that we are all in this together.

As Secretary of State Antony Blinken put it, "We've been through our own deeply painful ordeal with Covid. That's why we're so convinced that all countries need to work together to vaccinate the world—not in exchange for favors or political concessions, but for the simple reason that no country will be safe until all are safe. And all nations must transparently share data and samples—and provide access to experts—for new variants and emerging and re-emerging pathogens, to prevent the next pandemic even as we fight the current one."[74] That is what leadership is all about.

TAKEAWAYS

1. Honest, evidence-based political leadership is often the only defense against mis- and disinformation in a pandemic and therefore can have life-and-death consequences.
2. A pandemic is a force multiplier for economic and social inequality, so provisions should be planned ahead of time to lessen the effect.
3. It is better policy to subsidize businesses to keep workers employed than to subsidize those same workers for being laid off. That way, businesses remain stable and can gear up efficiently once the pandemic is over.
4. If economic safety systems are established and put in place before the crisis hits, government subsidies, safety nets, and other unusual expenditures will be far more efficient, less wasteful, and less likely to result in inflationary pressure.
5. Hospitals and healthcare systems must be prepared to transform into acute care–based operations in the event of a pandemic and to make difficult triage decisions when resources are inadequate, based on which patients have the best chance to recover. As Brad Spellberg puts it, "Everybody can have their own beliefs about how it should work, but if you don't get a coalition and consensus, making sure you're even more inclusive of ideas that you're not terribly comfortable with, you're never going to get anything done."[75]
6. The American healthcare system, as presently constituted, is in a poor position to scale up to meet the challenges of a pandemic. The fee-for-service model should be de-emphasized in favor of a value-based/population-based care model.

CHAPTER EIGHT

WHERE DO WE GO FROM HERE?

Those who cannot remember the past are condemned to repeat it.
—George Santayana, *The Life of Reason*, 1905

[Two Years After the Pandemic Ended]

Report of the Independent Commission on the SARS-3 Pandemic

Editor's Note

When it became clear after the last presidential election that there was no congressional or White House will to establish a commission to study the US government's response to the SARS-3 pandemic and make recommendations for future pandemic preparedness and response similar to what was done following the 9/11 terrorist attacks, several philanthropists representing a spectrum of political ideologies came together to fund and organize an independent commission to undertake this work.

Executive Summary

The three-year-long SARS-3 pandemic represents the single most devastating global public health event since the Black Death (bubonic and pneumonic plague caused by the *Yersinia pestis* bacterium) of the Middle Ages. Though the Black Death killed a greater percentage of the affected population at the time, in terms of sheer numbers, SARS-3 represents the most devastating public health event in human history. In the United States, 7,177,680 persons are confirmed to have died of the pandemic virus, representing 2.6 percent of the total US population. More than 36 million in the United States became ill enough to require standard or intensive care hospitalization. By way of comparison, the Great Influenza pandemic of 1918 resulted in approximately 675,000 American deaths, representing 0.64 percent of the population, while the Covid-19 pandemic was responsible for 1.1 million American deaths, 0.34 percent of the population. Globally, SARS-3 was responsible for more than 140 million reported deaths, compared with an estimated 50 million to 100 million for the 1918 influenza, and 8 million for Covid-19. It should be noted that actual numbers are likely substantially higher, due to incomplete reporting from overwhelmed state and local public health agencies in the United States and worldwide, as well as national and international underreporting for political reasons. Reasonable worldwide mortality estimates surpass 350 million. One sobering statistic is clear: Over the course of the SARS-3 pandemic, life expectancy in high-income countries decreased 6.8 years; significant decreases were also noted in low- and middle-income countries.

Despite the tragic numbers, the commission documented some positive and encouraging elements of the response that can be valuable in preparing for pandemics of the future.

Vaccine Development. Despite relatively rapid development of the mRNA-based vaccine platform during the Covid-19 pandemic, US coronavirus vaccine research lacked adequate support to move forward with broadly protective and durable vaccines once it was determined that the existing vaccines were subject to waning immunity profiles after a matter of months and, disappointingly, did not prevent viral transmission from those who had received vaccination. During SARS-3, an mRNA platform was employed to quickly produce targeted vaccines that were used as a stopgap while more durable vaccine candidates were being developed. Government support from the United States, the European Union, Japan, South Korea, and the Coalition for Epidemic Preparedness Innovations (CEPI) was sufficient to bring forward a broadly protective, live attenuated mucosal vaccine that was widely available in the United States and most Western countries in the third year, following impressive efforts for manufacture and distribution. However, the goal of providing these next-generation vaccines throughout Africa and the Global South fell short, partially due to failure to adopt an international pandemic treaty. Still, there is reason for optimism that recently established manufacturing facilities in Africa, India, and Indonesia will be able to meet the demand for mRNA-based coronavirus or influenza vaccines going forward, and the commission stresses that while this pandemic was caused by a coronavirus, the lack of a widely protective and durable influenza vaccine leaves the world highly vulnerable to the other most likely viral threat.

The commission emphasizes that many millions of lives could be saved, and the global economy protected, if the United States and other high-income nations commit to an ongoing program of development for vaccines offering robust protection, durability, and prevention of transmission for these most likely viruses of pandemic potential. But the commission notes this program must be accompanied by a coordinated effort to build a global industrial base and workforce to produce sufficient quantities of these vaccines to reach every country on earth in time to mitigate the course of whatever pandemic virus emerges....

———

In the third year of the SARS-3 pandemic, Dr. Adamu Kimani, director of the National Virology Reference Lab in Nairobi, Kenya, was asked by the World Health Organization to head up his country's vaccine-manufacturing program and facility. The project took advantage of the technology transfer hub WHO established in South Africa in 2021 to make vaccine technology available to fifteen sites in low- and middle-income countries throughout the world. With his extensive medical and laboratory experience, and strong leadership, Kimani was an excellent choice to head the effort.

One factor that made the Kenyan model work was Kimani's success in persuading governments to financially support manufacturing sustainability. He argued that for the facility to be able to scale up quickly in the event of an epidemic or pandemic, it had to be fully operational between pandemics and be permanently staffed by highly trained personnel. This meant continuously manufacturing seasonal influenza vaccine, produced through new cell-based technology. Thanks to Kimani's scientific diplomacy and fundraising acumen, the Kenyan government, other African nations, and WHO subsidized the costs of the facility and committed to purchase seasonal vaccines as they became available.

The facility could foresee cell-culture manufacturing within the next several years, which would allow it to scale up its output. Kimani was then able to spend time in the United States and India observing and studying public and private agencies and organizations to understand the development and manufacture of novel vaccines, with the intention of bringing the knowledge, training, and capacity back to Kenya.

...Respiratory Protection. Prior to the SARS-3 pandemic, the US government, together with three established manufacturers, launched a major initiative to develop reusable N95 respirator face coverings, washable up to fifty times. They became available during the second year of the pandemic. The adjustable fit of this respirator made it more comfortable and accommodating than previous N95 models, though it was estimated that less than 40 percent of the population used them consistently, mirroring challenges faced during Covid-19....

Dr. Peter Minekal sat on the deck of his Mission-style home high in the hills of Marin County, trying to enjoy the panoramic view of the Golden Gate Bridge and San Francisco Bay. But no matter how he tried to relax and clear his mind, his thoughts turned to that time nearly two years before, sitting in the same chair, when, without warning, he began to feel feverish. His chest was heavy, and he was having trouble taking deep breaths. His head started pounding and he felt weak. He went inside, hydrated himself, and lay down, but as the hours passed, he didn't feel any better. He knew that a bad flu could start this way, and he'd read accounts of people in 1919 feeling normal at the beginning of a crosstown streetcar ride and being near death by the end.

Could this be...? No, he wouldn't let himself think that way.

By the next day, he had a racking cough, a fever of 103 degrees, and a headache that felt like his skull was exploding. He'd had to cancel all his media appearances, which he hated more than almost anything. It galled him that his nemesis, that doom-and-gloom-predicting Dr. Robert Andrews of the

University of Michigan, was being asked to fill so many of the television slots. As bad as he felt, he kept telling himself he must have caught the seasonal flu from his daughter Michaela, who had brought it home during her semester break from Berkeley, where she was a junior. The university had decided to open for on-campus attendance the previous fall after a year of virtual classes. Michaela's symptoms were even worse than Minekal's, and so were her mother's, Minekal's wife, Vivian.

When Vivian began gasping for breath in the early evening, Minekal called 911. The EMTs arrived quickly and brought her to Marin General's emergency department, where the staff immediately administered a SARS-3 test, which was positive. She was put on supportive care and then transferred to the ICU. Although she was given antiviral drugs, the hospitalist, Dr. Erica Scott, told Minekal she was continuing to decline.

They'd asked if he had been or was currently ill, and he claimed that he had not been, marshaling all his strength to appear well.

"Have you or your wife been vaccinated against SARS-3?" Scott asked him.

"No," Minekal replied through the N95 respirator he was required to wear in the ICU.

"We're going to have to ask you to take a SARS-3 test," Scott said.

"No, I won't do that," Minekal declared.

"Then I'm going to have to ask you to leave right now, and please don't remove your mask until you are outside the building."

"Do you know who I am?" Minekal demanded.

"I don't care who you are, sir," Scott said, "and neither will hospital security."

Minekal turned abruptly to go. Between his brain fog and fatigue, he wasn't sure, but as he strode down the corridor, he thought he heard the hospitalist whisper to the nurse assigned to Vivian, "That's the doc who's been on all the television shows, saying that SARS-3 isn't a big deal."

By the time Vivian passed away later that week, with no family at her bedside, Minekal was beginning to recover, but Michaela was still sick and couldn't attend her mother's funeral. Clips of Minekal's recent media appearances, intercut with images of him at the cemetery as Vivian's coffin was lowered into the ground, defiantly unmasked but clearly grief-stricken, were shown frequently on cable news channels and went viral on the internet.

Since then, he'd ceased all his media appearances and pulled deep into himself. Was it the grief over Vivian's death, or was there more to it? *he wondered.*

...Economic Impact. SARS-3 has had the greatest negative impact on the global economy in modern time. While it is still too early to evaluate the far-reaching effects, we have seen the plunge in world securities and commodities markets lead to a catastrophic international recession, for which recovery is still not on the horizon. Supply chains slowed dramatically or halted altogether. Hunger, job loss, and homelessness increased steeply worldwide. Even in the United States, reported homelessness reached its highest level since the Great Depression of the 1930s, despite government efforts at emergency rental assistance, stimulus payments, federal aid to state and local governments, and a temporary eviction moratorium similar to that instituted during Covid-19.

Global mitigation efforts for illnesses such as malaria, tuberculosis, polio, and childhood diarrheal diseases plummeted, with severe consequences for national health and worker availability. Metrics on economic retrenchment in low- and middle-income countries are contained in the body of this report, but economists for the five regional United Nations Economic Commissions have estimated that it will take at least a generation for these countries to recover to prepandemic levels.

...Family Dislocation. The SARS-3 pandemic had a devastating effect on family structure, akin to that observed from major wars. As is clear from the WHO's excess mortality-derived estimates presented within this report, millions of children around the world were orphaned through the death of parents, grandparents, or other caregivers. With grandparents taking over care, many children face the possibility of being left orphaned a second time, since grandparents are often in the highest-risk cohort for serious disease or death.

The downstream effects of all of this — short-, medium-, and long-term — include, but are not limited to, increased levels of institutionalization; physical, emotional, and sexual abuse; traumatic grief; other mental health issues; adolescent pregnancy; poor educational outcomes; increased incidence of poverty and homelessness; and chronic and infectious disease. The greatest numbers occurred in Africa and Southeast Asia, though no region was spared. In many instances, little has been done for these children.

It is essential for governments not only to address this desperate situation now but to incorporate plans, both internationally and on a countrywide basis, to deal with children left without parents and care-takers in a future pandemic. This includes ensuring vaccine availabil-ity throughout the low- and middle-income countries, as well as instituting secondary measures such as economic support, alterna-tive care options through safe and loving family-based care support services, physical protection, ready school access, and health out-reach. High-burden countries in sub-Saharan Africa will require spe-cial attention...

It had taken many weeks, but Abdirahim Ali Salat, the immigrant who left Somalia for Minnesota at the very beginning of the pandemic, finally recov-ered from his severe case of SARS-3 but was too weak from the extended effects of the virus to pursue a tech job like his brother Abshir's. His own suffering was overshadowed when he learned from Abshir that while Abdirahim had been in the hospital, his wife, Bishaaro, and their children had gotten sick. Abshir waited until he was stronger before he told Abdirahim the awful truth: His entire family had perished. Neither brother had been able to get in touch with any family in Somalia for weeks. Abdirahim longed to return to learn more about what had happened, but without a job or money, that would have been impossible even if flights from Minneapolis home were still consistently running. He hoped for the best, grateful to Abshir and his wife, Caaliyah, for making him feel welcome and letting him stay with them and their son and daughter as long as he needed to.

Abdirahim had planned to seek work from the company that employed his brother, but just as he was feeling well enough to start the application process, Abshir came home despondent with the news that the company was laying off workers and that he had lost his job. They'd lost too much business during the pandemic to keep everyone employed.

Not only was Abshir devastated to be out of work; that night he started coughing and spiked a fever. A home test was positive for SARS-3, which soon spread to Caaliyah and the children. They'd all received the initial vaccination as soon as they could, but they had not gotten any boosters. Abdirahim apparently still had immunity, for he did not become ill even as the rest of the family grew sicker.

The children recovered within a week or so, but Abshir and Caaliyah took a downward turn. Abdirahim took them both to the same emergency department where Abshir had taken him when he first arrived. He remembered the kindly Dr. Thomas who was in charge. A different doctor was in charge now, who said Dr. Thomas was out sick. He told Abdirahim that his brother and sister-in-law both belonged in the ICU, but since it was more than full, they would be placed in a temporary intensive unit that was set up in the hospital's parking lot.

Two days later, he was home making breakfast for the children when he answered his cell phone. It was a nurse who sounded like she was trying hard not to cry. Abshir and Caaliyah had both died during the night. He struggled to keep his composure as he tried to figure out how he was going to break this awful news to his niece and nephew. Beyond that, he wondered how he could care for them without a job or any way to pay the rent.

In the months that followed, he took any odd job he could find, including making deliveries on a bicycle for local restaurants and grocery stores. It didn't pay much, but he was able to take home leftover and discarded food, which he shared with the neighbor who'd been a dear friend of his brother and sister-in-law's and was helping out by watching the kids when Abdirahim went to work. With the pandemic finally over, he wondered if the employment situation would improve, or if he should go back to Somalia or what used to be a better prospect, Kenya, but with his own wife and children gone, the thought of being back where he'd had such a happy life was overwhelming to consider. Besides, who would take care of Abshir and Caaliyah's children? He'd come here imagining such a different life for them all.

...Education. Once the severity of the pandemic was established, the question of whether or not to close K–12 schools, as well as higher education — and if so, for how long — became a point of controversy and conflict in communities throughout the United States. The issue was framed on one side in terms of protecting students, teachers, and administrative and support staff and avoiding having children bring the virus into often multigenerational households, infecting parents, grandparents, and immunocompromised individuals being cared for at home. And, as a practical matter, it was difficult to keep schools open, since so many teachers, administrators, and support staff had died, become severely ill, or suffered the effects of long SARS-3.

On the other side was the concern that students could lose years of learning, as well as having existing achievement and skills erode. This was further complicated by the need to provide care for children who remained at home during closings, which inordinately burdened lower-income and working-class families. The resolution to these issues varied tremendously among the states, leading to a crisis of faith in schools that was never fully resolved.

With the benefit of hindsight, it appears that guidance from national and local authorities was diametrically the opposite of what public health logic would have suggested. In the first year of the pandemic, when many jurisdictions closed schools for the entire academic year, there was a lower level of transmission to and from children compared with years two and three, similar to our experience with SARS-CoV-2. By then, variants increased infection, transmission, and severity of illness in children. But few communities tied their operations to local levels of infection, which were largely blamed on overwhelmed public health agencies that were unable to provide accurate surveillance data on levels of infection. And most localities could not long withstand public demand to reopen schools. In some communities there was a notable increase in serious outcomes for custodial, cafeteria, and maintenance staff, as well as school bus drivers, possibly a result of not being able to access healthcare resources.

Remote-learning initiatives highlighted economic disparities, with less-affluent communities possessing fewer physical and parental resources to help their children navigate a digital learning environment. In many cases, this led to markedly divergent levels of academic achievement and setback among affluent and economically challenged communities...

Hardly a day went by during the height of the pandemic when CDC director Tamara Goldfield hadn't received at least one email, text, or letter from a school superintendent or teachers union official with conflicting complaints and demands.

What was the CDC's guidance on this? they all seemed to be asking.

Unfortunately, there was no simple or uniform answer. It depended so much on the facts on the ground at each location; the density of the living situations of the residents; what was happening with the local healthcare systems; the levels of infection within the community and how long they had been that way; and the political climate of the state.

Many parents, except for a certain cohort of the well-off, wanted classrooms to open or reopen. Principals mainly were in favor of in-school learning but were afraid of liability or political consequences if anyone became seriously ill as a result. The teachers' unions wanted to keep schools closed and said they should not be "sacrificed" for political expediency or the convenience of parents. Where infection rates were particularly high, hospitals were asking for temporary closing of schools and most public gatherings, just to give them breathing room to flatten the pandemic curve and even out their patient loads. All in all, it was an impossible mess that there was no way to resolve to everyone's satisfaction.

Goldfield and her husband were personally fortunate that the elementary and middle schools their daughter and son attended in the Buckhead section of Atlanta were academically high achieving, with administrators who were open to informed advice, and an annual budget that included tablets or laptops for most students. That wasn't universally true throughout the South, or much of the rest of the country, for that matter. She didn't want to get involved influencing her kids' schools — the optics wouldn't be good — but she was afraid it was only a matter of time before the whole thing blew.

Relatively speaking, she had to conclude that her kids had come through the pandemic experience better than most. But the standardized testing was hard to dispute: They and their classmates had definitely lost grade-level progress. Whether or when they would all catch up was impossible to know.

...*Messaging.* The public messaging presented a mixed picture. Both the World Health Organization and the Centers for Disease Control and Prevention were unclear throughout the pandemic on whether viral transmission was primarily airborne, despite compelling evidence. This led to confusion regarding the most significant means of disease mitigation.

After a slow start, the White House messaging improved dramatically when the president undertook a series of "Fireside Chats," in the manner of President Franklin Roosevelt, explaining, "Science is not truth, but rather a search for truth," and that information and recommendations would evolve and be updated as the science of the virus itself evolved. This substantially increased the general level of trust among many demographics, as did his public vaccination along with the four living former presidents, representing both political parties. And the president conveyed humility, admitting what remained unknown about the virus as it mutated, for example, and empathy by relating to what ordinary people were going through.

But conditions changed after the presidential election, when the other party came into power and the new chief executive implied that a change in national attitude could have prevented the worst effects, and claimed that the pandemic was essentially over despite the continuing statistics to the contrary....

––––––––

While the president received generally favorable marks for his communication and handling of the pandemic, pollsters explained the election results as "pandemic fatigue" and said the American public simply wanted a change. They likened it to Prime Minister Winston Churchill's Conservative Party defeat in 1945 after leading Britain through to victory in World War II.

In his inaugural address, the new president called on the nation "to turn over a new leaf and put this unfortunate experience behind us. Though the pandemic was unquestionably devastating, for too long we let it control us, rather than taking charge and determining our own fate. We let far too much of our lives shut down and spent far too much money on dubious relief efforts that have left us in the inflationary and debt-ridden situation in which we now find ourselves.

"The people have spoken. With this leadership change, it is now time to declare the pandemic under control and to move boldly forward into a hopeful future. We learned, many times to our own disadvantage, that so-called scientific experts didn't have all the answers and didn't have a monopoly on truth. It turned out that good old common sense was our greatest strength.

"As a first step in bringing back fiscal sanity, we will be cutting the budgets of the NIH and the CDC and returning research and development to the private sector, where it belongs in a free market economy like ours."

...Healthcare Systems. The lack of a comprehensive medical system available to all failed the United States. The numbers of healthcare workers and hospital beds were grossly inadequate to meet exigent needs during peaks of infection. This statement takes nothing away from the heroic efforts of first responders, physicians, nurses, technicians, other healthcare professionals, and the wide variety of support staff. However, many times, and in many locations during the pandemic, they were simply overwhelmed.

As comprehensively discussed in this report, the United States, like most high-income countries, faces a critical decision as it attempts to prepare for future epidemics and pandemics: Are we willing to devote the resources to develop reserve capacity — both physically in hospital facilities and in staffing — to the existing healthcare system? The fundamental question is whether we can accept the costs of these resources when we do not need them so that they are available when we do.

The pandemic exacerbated the existing crisis in mental health services in the United States and throughout the world. At the same time that the vulnerable lost their emotional support systems due to physical loss or incapacity of close friends, relatives, and therapists, hospitals and healthcare facilities could not handle the influx of new cases of those seeking help. Many emergency departments, already in triage or semi-triage mode due to SARS-3 cases, were overwhelmed with those in mental health crisis, with neither the staff nor the facilities to deal with them. As with the Covid-19 pandemic, we saw a virtual breakdown in the mental health support system. The dramatically increased rates of suicides, drug overdoses, and public disturbances speak to the severity of the problem...

Erin Thomas had one doctor on her staff who had served in the army in Iraq and Afghanistan. He told her that what they'd faced in the emergency department was just as intense and overwhelming as what he'd dealt with overseas. And like combat casualty work, the onslaught came in waves. In the battle space, the waves came in every time there was a new offensive, skirmish, or attack. In the hospital, it was every time a new variant emerged and local infection spiked.

It was the first time in her career that Thomas actually had to institute triage—a medical-allocation decision process, usually considered in acute battlefield or civilian mass-casualty events. It troubled her to do so, but dire conditions required her to withhold treatment from those unlikely to survive in order to have the resources available for those patients with a better chance of survival. People in critical condition—suspected heart attacks and strokes, those suffering blood loss and head traumas—were categorized as either likely to make it with available treatment or not likely to make it. Those with good chances got priority, but all non-life-threatening cases had to be turned away. Knowing that anyone who had a limited chance to survive could not be a priority for care violated everything Thomas believed in, but it was the only way they could manage.

They were short of everything: staff, supplies, medicines, even the most basic necessities, like sterile water—the nasal cannulas so many patients needed to breathe were being used faster than they could be restocked. Her hospital frequently ran out of IV bags, making administration of many critical drugs extremely difficult. The situation was desperate.

It was even worse in the ICU, whose director was instituting the same dreaded triage approach Thomas and her team were employing in the emergency department. The ICU had patients on CRRT, chronic renal replacement treatment, whose kidneys had completely stopped working and who were too sick to tolerate the standard three- or four-hour dialysis three times a week. Instead, CRRT was a continuous, twenty-four-hour process, and it required two nurses to staff each patient. The ICU director lamented to Thomas, "We just don't have critical ICU-qualified nurses to put four patients on CRRT who are not going to survive. I had to tell the renal docs we couldn't take any more of their cases. Otherwise, folks who have a chance at survival aren't going to get it. We have to choose who lives and who dies."

On top of the demoralizing day-to-day decisions, surrounded by so much death and loss, Thomas was still weak and spent from recovering from her own case of SARS-3. She'd gone back to work as soon as she was cleared, gritting her teeth and trying to power through each day. But she found she had trouble concentrating and never regained her strength. Eventually, she was diagnosed with long SARS-3, and the head of the infectious disease department said she was not fit for work.

As weeks, and then months, dragged on, and then the first years post-pandemic, Thomas was only marginally better. She didn't know if she'd ever be able to go back.

IN THE PENULTIMATE SEQUENCE of Charles Dickens's classic *A Christmas Carol,* the newly chastened Ebenezer Scrooge, brought to an overgrown churchyard by the Ghost of Christmas Yet to Come, stares at the gravestone bearing his name and plaintively asks the ghost, "Are these the shadows of things that Will be, or are they shadows of things that May be, only?"

Scrooge had the good sense and good fortune, after his harrowing Christmas Eve, to heed the spirit's warning, turn his life around, and positively affect the lives of those around him. When it comes to preparing for the Big One, we may not be as fortunate, unless we heed the timely warnings we have been given. And so let us review the relevant points, which we hope have been brought home in our thought experiment.

PANDEMICS ARE NOT OPTIONAL

The esteemed and decidedly not overly optimistic Bipartisan Commission on Biodefense stated in its 2024 *National Blueprint for Biodefense,*

> We...believe that the United States, its allies, and partners in industry, academia, and nongovernmental organizations can eliminate pandemics entirely in 10 years by fully implementing the recommendations we made in our earlier report, *The Apollo Program for Biodefense.* Ending pandemics is more achievable today than landing on the Moon was in 1961. Our Nation has the right stuff. We can do what it takes—if we want to.[1]

Much as we would love to agree with this laudable sentiment, it doesn't have much basis in reality. The Big One is not optional; it's not an if but a when. As we noted early on, we must understand that we have no way of avoiding pandemics.

Yes, it was practically unthinkable to land humans on the moon in 1961 when Navy Commander Alan B. Shepard Jr. climbed into the Mercury Friendship 7 capsule for the first fifteen-minute American suborbital space flight. But by 1969, that goal had been achieved. However, even if the United States and other high-income nations were willing to commit a proportionate share of national resources as we did to the space program in the 1960s, we must distinguish between what was essentially a complex engineering challenge that did not require international cooperation and the far more complicated scientific puzzle of creating, in advance, universally applicable, highly effective, and durable vaccines that would halt transmission against two or more airborne-transmitted viruses of pandemic potential. If science could achieve that, with the additional complex challenges of rapidly scaling up manufacturing capacity, establishing a global distribution network, achieving worldwide foreign government buy-ins, and convincing a significant percentage of the population to accept these—presumably government-paid-for— vaccines, then we might possibly be on the verge of eliminating airborne-transmitted viral pandemics. Of all the complex crises the world faces, with climate change at or near the top of the list, this one is among the most solvable.

But we are nowhere near that point, in terms of both financial commitment and scientific progress. Despite the impressive strides, vaccines for influenza are no more effective now than they were in the 1950s. Covid vaccines, while new and innovative, are no more effective than flu vaccines. This state of affairs will do little to overcome the current vaccine hesitancy and anti-vaccine culture. Forty to 50 percent protection is better than zero, but in the eyes of many, these vaccines will hardly be worth getting. So, while being prepared to deal with current realities, we need to focus on developing more effective vaccines.

As we have demonstrated, it is unlikely that we would even know of an outbreak of pandemic potential before it had spread beyond an international border. The global commercial aviation system is far more robust than the global disease surveillance system. After SARS was transported from rural China to Hong Kong in 2003, it spread to five countries within a few days.[2] At the beginning of Covid, the SARS-CoV-2 virus was all over northern Italy and had insinuated itself into the United States from there while we were still trying to decide whether plane flights from China should be curtailed.

We support the thirty-six recommendations and 185 action items detailed in the *National Blueprint for Biodefense.* These will help us prepare. But what I wrote in my 2005 article in *Foreign Affairs,* "Preparing for the Next Pandemic," remains true two decades later: "The reality of a coming pandemic, however, cannot be avoided. Only its impact can be lessened."[3] Our goal should not be to try to make pandemics "optional" but to be ready and prepared with research, strategies, drugs, and nonpharmaceutical interventions to mitigate the pandemic when it emerges.

Among all the things that went wrong during the Covid pandemic response, some accomplishments can be seen as great successes. A lifetime of lessons were learned in just the first two years regarding intensive care clinical practice. Time bought significant improvements in patient management, initially learned through trial and error in treating millions of patients and eventually through the results of large randomized clinical trials. The mRNA vaccine technology was also a major success, even with all the problems getting the public vaccinated and the relatively modest duration of protection against serious illness following vaccination. Millions of people worldwide survived the pandemic because of these vaccines.

In the next pandemic, while mRNA vaccines may not provide ultimate and durable protection, they can serve two important purposes. As we've noted, since they can be manufactured relatively quickly in large volumes compared with other platforms, they could act as a stopgap against the virus while more effective and durable vaccine candidates are developed. No less an authority than Dr. Stanley Plotkin, emeritus professor at the University of Pennsylvania, a mentor to me and a giant of vaccinology who played pivotal roles in the development of rabies and rubella vaccines, suggests that the mRNA vaccine could be a primer, followed up by a mucosal vaccine whenever it becomes available, perhaps six months or a year later. Such a vaccine combination could hasten the end of the pandemic.

PANDEMICS CAN LAST NOT DAYS OR WEEKS BUT YEARS

This is how the typical pandemic disaster movie unfolds: Some intrepid family or emergency physician realizes an uncommon pattern of disease that the rest of the medical establishment has missed. He or she has trouble convincing the powers that be, but soon every country in the world is reeling and hospitals are overwhelmed— from big Western cities, where public health takes over sports arenas and sets up tents in parking lots to care for the sick, to rural African villages, where the entire community gamely pitches in while the one Western volunteer says, "I wish I saw this kind of cooperation back home."

Amid all the suffering and death, a small band of plucky researchers are burning the midnight oil in a lab somewhere, mixing all kinds of chemicals in test tubes and holding them up to the light, with hopeful looks on their faces. Finally, after weeks of round-the-clock toil, one of them declares, "I think I have it!" While the others look on, he injects it into the arm of a colleague near death from the virus, with whom he is secretly in love. They all watch breathlessly, hoping against hope, and within minutes, the beloved colleague has miraculously begun to recover. The next day, thousands of gallons of the new vaccine are rushed out around the globe, world leaders breathe a collective sigh of relief, and the pandemic is declared over.

It doesn't work that way in real life.

The 1918 Great Influenza lasted more than three years. Covid-19, so labeled because it was first identified in 2019,[4] stretched past 2023, and its effects will last far beyond that. Even under the best of circumstances, coming up with a vaccine against the Big One will take many months, rather than a few weeks; and manufacturing sufficient volume and getting it into arms or noses throughout the world following Emergency Use Authorization will take months more.

FLEXIBILITY IS KEY

We must also understand that the situation will not remain static throughout the pandemic. If it is caused by a coronavirus, we should expect periodic variants that can make the virus either more or less transmissible and more or less virulent. And that level of transmissibility and virulence will not necessarily portend what will happen as these viruses continue to mutate. As influenza or coronavirus strains evolve during the course of a pandemic, we are likely to see surges in cases over a several-month period. This is why plans made, and information and guidance circulated, early in the pandemic will be subject to ongoing updating and reassessment as the multitude of relevant situations and data develop. This is not flip-flopping, it is science. Flexibility is key.

PANDEMICS AFFECT EVERYONE

Even if you are lucky enough not to contract the airborne virus, someone you know and care about most likely will. But even beyond that, a pandemic would so severely affect the global supply chain that both ordinary and durable goods, food, medicine, and the staples of everyday life would be in short supply or not available. There would be major shortages in all countries of a wide range of commodities, not only food, but also soap, paper, light bulbs, and gasoline, as well as parts for cars, airplanes, trains, military equipment, municipal water pumps, and electrical generation plants. Even coffins to bury the dead would be in short supply. With Covid, we saw how connected the world's economies are.

The message here: *When it comes to fighting microbes, America First only goes so far.* In the United States, most of our critical and, in many cases, lifesaving generic drugs come from China and India,[5] both of which would be prime targets for viral spread, resulting in shutdown of manufacturing plants. We have been advocating for years for this type of pharmaceutical manufacturing to be reestablished in the

United States and other countries we can count on, as a matter of national security. But that would necessarily involve some form of government subsidy, since the profit margin on most generics is extremely thin, and even overseas, companies are getting out of the business. This means that as consolidation in China and India has occurred, it has created a gaping vulnerability for the United States and the Western world.

The truism that no one is completely safe until everyone is safe is a truism because it happens to be true. In the words of the late Nobel laureate Dr. Joshua Lederberg, whom we quoted at the beginning of Chapter One, "Bacteria and viruses know nothing of national sovereignties.... No matter how selfish our motives, we can no longer be indifferent to the suffering of others. The microbe that felled one child in a distant continent yesterday can reach yours today and seed a global pandemic tomorrow."

Or, as the poet John Donne wrote, "Send not to know for whom the bell tolls; it tolls for thee."

Accordingly, in preparing for the Big One, we must not let the same thing happen that occurred with Covid, where high-income nations ended up with plenty of vaccine—often more than they could use—while low- and middle-income countries had very little, despite COVAX's good intentions. Not only must we develop new and effective vaccines; we must also, by international agreement and cooperation, plan for a means to scale up manufacturing to meet global need, along with an efficient system to transport and distribute them, even if a cold chain requirement is involved. We will need an international approach to public funding that will pay for the excess production capacity required during a pandemic.

Unlike in many fields these days, ethics remains a vital and integral component of medicine and public health, and thus there is compelling reason to regard the rest of the world with the same compassion and empathy we feel for our own people. But on a practical level, there is nothing particularly altruistic about sharing vaccine with low- and middle-income countries in sufficient quantities to protect their populations. It is simply self-interest. Now that the globe can be circumnavigated in less than forty-eight hours, distance provides no protection from infectious diseases. While someone in a remote village in the Western Pacific or sub-Saharan Africa is sick with a novel airborne respiratory virus, people on the other side of the world may be in imminent danger, a fundamental fact of nature in our modern world.

We realize how unlikely this level of global cooperation is in reality, given the state of international relations and each country's natural tendency to keep critical drugs and vaccines for its own people. That probability, however, doesn't make this any less important. Manufacturing countries must have the capability and capacity to turn out vaccine stocks for the rest of the world, and there should be international dialogue and planning for the mechanics of how vaccine stocks would be allocated.

Even in the United States, there will not be sufficient antivirals to meet the need for at least several months. Assuming effective antivirals even exist for whatever the pandemic virus turns out to be, we will have to figure out who gets priority among those who are seriously ill. Healthcare workers and first responders? Political and business leaders? The elderly and immune-compromised? Essential workers and drivers? Each cohort will have its advocates. It is far better to struggle with the ethical issues involved in determining such priorities now, in a public forum, than to wait until the crisis occurs.

Another issue is that while SARS-CoV-2 primarily affected the elderly and immune-compromised with severe disease, that wouldn't necessarily be true of the next pandemic. Keep in mind that in the 1918 influenza, more than half the people killed were eighteen to forty years old and largely healthy. These deaths were likely caused by a virus-induced response of the victim's immune system—a cytokine storm,[6] as we described in Chapter Four—that led to acute respiratory distress syndrome (ARDS). In other words, in the process of fighting the disease, these healthy individuals' robust immune systems overreacted, severely damaging the lungs and resulting in death. Today, the medical establishment around the world is not much better prepared to treat tens of millions of cases of ARDS than it was more than a century ago.

And even though the SARS coronavirus, for example, infected only about 8,000 people in 2003 before it was brought to a halt, about 10 percent of them died,[7] showing that our thought experiment for SARS-3 is not far-fetched.

ANOTHER "LONG" SYNDROME IS A DEFINITE POSSIBILITY

Just as SARS-CoV-2 produced the still only partially understood condition known as long Covid, we must anticipate that a future viral pandemic could trigger another set of specific or ill-defined symptoms lasting weeks, months, or years. As with long Covid, the personal, medical, and societal burden of such a development could be devastating. Therefore, as part of our planning for the Big One, we will need enhanced research, medical countermeasures, and resources in place for adequate care.

Further investigation into the specific triggers and appropriate treatment for long Covid will help prepare for the possibility of "long Disease X," whatever that turns out to be, and how to deal with it. And if, by the time the Big One hits, we have developed more effective antivirals and vaccines for both influenza and coronavirus, we may be able to short-circuit any long syndrome before it becomes a widespread phenomenon. At the same time, we should anticipate the likely healthcare needs that would accompany a large population of patients needing extended care. Our hospital and healthcare surge capacity, as well as that of other high-income countries, is virtually nonexistent as the Covid pandemic demonstrated. We always consider surge capacity in strategic defense planning. It would seem to be simple common sense that we would do the same for strategic pandemic planning.

From an economic and societal standpoint, in addition to the cost of caring for the sick, we must take into account the effect of potentially losing a sizable percentage of our workforce, from those immediately incapacitated (or worse) to those who may be partially or completely disabled by their long Disease X illness. We saw some of these effects during Covid, which hit healthcare particularly hard through a combination of short sick leaves, longer incapacity due to lasting effects of the virus, and professional burnout. No industry is likely to be spared in the Big One. Large companies and government contractors often have disaster and succession plans in place, but they will suffer nonetheless if they need to replace personnel who have specialized training and expertise who are out for extended periods or cannot return to work. The effect will be felt hardest in small businesses, which, as we saw during Covid, may well be unable to absorb the impact of prolonged absences. If some of these businesses fail and remaining employees are let go, the economy will feel the one-two punch of high unemployment and the need to support those sickened and disabled by the pandemic disease.

Our pandemic planning needs to account for all these risks.

PANDEMIC PLANNING IS LIKE BUILDING AN AIRCRAFT CARRIER

Both are critical to national defense. Building on the theme we ended with in the previous section, biological security is just as important as military security, and the United States and like-minded countries need to accept the idea that it means going to war against a microbial enemy potentially far more dangerous than any conceivable human foe.

What the defense establishment has been able to master, but the public health community has not, is making the congressional funding apparatus understand that weapons systems cannot be developed or built on a year-to-year budgeting basis. You can't build an aircraft carrier, a new fighter jet, or any other sophisticated weapon system by negotiating and requiring budget approval every two years. And if that carrier or weapon system is not available at the beginning of an armed conflict, the consequences will speak for themselves.

Likewise, pandemic preparedness is a long-term investment, incorporating surveillance, antiviral and vaccine development, personal protective equipment manufacturing and stockpiling, non-pharmaceutical interventions, planning for economic safety nets, etc. The investment must be consistent and continuously built. As an example, it would be difficult, if not impossible, to build a reporting system on the fly during a pandemic, as we saw during Covid. If one exists, it cannot be allowed to lapse or fall into disuse, so that we must start from scratch when the next pandemic hits.

PANDEMIC PREPAREDNESS IS A WISE FINANCIAL INVESTMENT

Shortly after Donald Trump assumed the presidency for the second time, he and his newly appointed HHS secretary, Robert F. Kennedy Jr., began cutting critical funding and personnel slots throughout the government health establishment, including the NIH, CDC, and FDA. We consider this an extremely shortsighted policy. Just as we have maintained that the results of dollars spent on research, planning, and prevention end up saving many times the investment in positive outcomes, the investment not made is likely to result in negative consequences far in excess of the dollars supposedly saved.

Because so much of state and local public health funding is funneled through the federal government, when resources are dramatically reduced, that reduction in support means there is real danger of missing early signs of an outbreak when it could most efficiently be contained or mitigated.

We are not the only ones shouting this theme from the rooftops.

Vanessa Kerry, MD, CEO of Seed Global Health, which helps provide nursing and medical training support in resource-limited countries, stated it well in an opinion piece in *STAT News:*

> Annual funding of $31 billion for pandemic preparedness would save trillions in the kinds of losses experienced during the Covid-19 pandemic, and which occur annually on smaller scales with cholera, Ebola, and other outbreaks. Weak health systems in turn can often drive migration across borders, putting pressure on the host country's resources. Health is a fundamental prerequisite for stable communities.[8]

In a 2023 *Washington Post* op-ed in support of a global pandemic fund, Harvard economist Lawrence Summers and coauthors Robert Hecht and Shan Soe-Lin wrote,

> Experts predict a 50/50 likelihood that we will see another pandemic as deadly as covid-19 or greater over the next two decades. Given that covid-19 has cost the global economy $12.5 trillion (and climbing), $20 billion a year or more on pandemic preparedness would be a sound investment, even if it lowers the chances of more losses by just 10 percent.[9]

Aside from the dollars not being invested, there is the danger of cutting back on the expert workforce involved in medical research and outbreak readiness, as seen in the early months of the second Trump administration, especially at the NIH and the CDC. We acknowledge that any presidential administration has a laudable interest in trying to cut back government waste and duplication, but potential cuts should be carefully scrutinized as to their likely impact on an agency's ability to effectively perform its established function. The sweeping and unprecedented reductions in force demanded by Trump's Department of Government Efficiency (DOGE) leader, Elon Musk, have appeared rash, imprudent, and broad-brushed rather than strategic—cuts made with a machete rather than a scalpel. The ultimate impact of these cuts and actions during the first one hundred days of the administration is unclear; however, it is likely to be devastating.

Many of the experts who are needed in order to respond to future pandemics have had their career paths obliterated by funding cuts in the first months of President Trump's second tenure. This could have the effect of severely compromising the next generation of public health expertise and effectiveness, which the country cannot afford.

Even as we write this, we are closely monitoring the evolution of an H5N1 avian influenza virus, already found in more than fifty animal species, notably poultry and dairy cattle. In addition, a novel coronavirus in China carried by bats could become a more serious threat than SARS-CoV-2. It could spill over any day, or perhaps never, but we wouldn't want to bet our lives on it by not having the proper experts on board to deal with it. And no matter when you're reading this, there will be threats the public health community is watching for signs of outbreak. The pandemic clock is ticking. We just don't know what time it is.

SCIENCE-RESISTANT LEADERSHIP IS DANGEROUS

While I have always advocated humility and caution in conveying what we know and don't yet know in emerging science, I can say with certainty that any political leadership that undermines faith in science is dangerous, and in the United States, this is true of every branch of government—executive, legislative, and judicial—and at all levels, from the president down through governors and state legislators to local commissioners, school board members, and the like.

This is true not only in the United States, but throughout the world. Consider South African president Thabo Mbeki's late 1990s and early 2000s challenging of the fact that AIDS was caused by HIV and his leading a policy denying antiretroviral drugs, which he characterized as poisons, to AIDS patients. This action was estimated to have led to more than 300,000 preventable deaths.[10]

In the United States, this has been most evident in undermining confidence in vaccines. When, shortly after the 2024 presidential election, president-elect Donald Trump nominated Robert F. Kennedy Jr. to be secretary of health and human services, Kennedy, a known vaccine skeptic and promoter of the repeatedly debunked claim that vaccines can cause autism, said, "We're not going to take vaccines away from anybody," before stating, "I'm going to make sure scientific safety studies and efficacy are out there, and people can make individual assessments."[11] He may have been emboldened by Trump's declaration that he would allow Kennedy to "go wild on health."[12]

Well, the fact of the matter is that while all vaccines, and all

medications, carry slight risks, the data on all vaccine safety and efficacy studies is already out there, and the scientific conclusions are
clear. Smallpox, the scourge of history, has been eliminated from
the face of the earth by vaccination. Polio, which sent terror into the
hearts of parents for centuries, has been eliminated in all but a few
countries. And diphtheria, once called the "strangling angel of children," has been largely consigned to the dustbin of history thanks
to vaccination. A 2024 study detailed that for children born in the
United States between 1994 and 2023, nine vaccines prevented an
estimated 1.1 million deaths, 32 million hospitalizations, and 508 million cases of illness, saving $540 billion in direct medical costs
and $2.7 trillion in indirect costs.[13] And every time a skeptic cites an
adverse reaction to a vaccination, such as myocarditis in teens
(which is generally mild and brief) resulting from a Covid shot, they
fail to cite the statistics showing that such outcomes are far more
likely to occur from Covid itself.

Though it may sound reasonable to say that people can make
their own individual assessments, when leadership undermines faith
in vaccines or any other scientific principle, and at the same time
promotes unproven or discredited treatments, as Trump did in his
first term with ivermectin and hydroxychloroquine, the health of
the nation — particularly in the midst of a pandemic — is going to be
severely compromised.

We are also concerned that if skeptics of established science are
placed in positions of government leadership and allowed to appoint
like-minded people to critical positions such as the Advisory Committee on Immunization Practice, the already fragile trust in science
exhibited by a sizable portion of the public will be further eroded,
with potentially dire consequences to the public's health and well-
being.

We have been candid about the public health establishment's
mistakes and failures of communication during the Covid pandemic. But regardless of how much vaccine and medical countermeasure progress has been made when the Big One hits, a robust
and effective response will only be possible if public health officials

and government leaders work together for a common goal, admitting what we don't know, assuring the public that we are working toward learning more, and not calling into question the basic tenets of our fight against dangerous microbial enemies.

FINAL THOUGHTS

We've noted that when we are asked what the ordinary citizen can do to prepare for a pandemic, there often isn't a great deal to suggest, other than investigating the health- and science-related records of candidates for public office and advocating with elected officials already in office. However, once a pandemic actually hits, the success of the containment and mitigation efforts relies substantially on the actions and compliance of ordinary citizens. That is why we find it unreasonable that so many assert personal choice and freedom in resisting actions like wearing N95 respirators and accepting vaccination. We would hope that people could rely on one another to take whatever measures they can to protect themselves, their loved ones, and those around them in the larger society. This is where the example and ethos of the Scandinavian countries is so powerful.

The only currency public health has is trust, and we have seen how far that has been eroded during Covid-19. But public trust is a two-way street. We want the citizenry to trust the government and health officials on pandemic mitigation measures, but we recognize that those officials must continually earn that trust. For example, if a government agency tells you that you will be protected by wearing a face mask such as a surgical mask, and you later learn that you are only truly protected with an N95-grade respirator, that agency has lost your trust. It's perfectly reasonable that people want the government to protect them from foreign enemies and terrorists, even though the risk of invasion or mass terrorism is actually quite small. It should be just as reasonable that they would want the government to protect them from dangerous microbes, where the risk is actually great. That is a key aspect of the social contract, but again, only if trust is earned. What happens if it hasn't been recovered by the time the Big One hits?

The outcomes of our ongoing thought experiment of an imagined SARS-3 pandemic, presented in the opening sections of this final chapter, are neither as good as we could hope nor as bad as we might fear. But as we have made clear, hope is not a strategy, and fear is not an excuse. As with Scrooge, we have seen what's coming. The choice of how we prepare for and face it is up to us.

ACKNOWLEDGMENTS

Though our two names are on the cover, this book rightfully has three authors. Ann Hennigan Grace has worked with us since the beginning of the project, researching, organizing, editing, and fact-checking every aspect of the narrative. She has also kept us on track throughout the many iterations of each chapter as new developments in the Covid pandemic and ongoing increasing knowledge in coronavirus research presented themselves. It is no exaggeration to state that we could not have completed the book without her, and her influence is on every page.

As always, we are grateful to our intrepid agent Frank Weimann of Folio Literary Management for his unflagging enthusiasm, support, and guidance, and active involvement with our work.

We have been fortunate to have two of the best editors in the business on this book. Tracy Behar, who edited our previous book, *Deadliest Enemy: Our War against Killer Germs,* saw what we wanted to do and partnered with us to shape the concept and figure out the best way to tell the complex story. Then, when Tracy moved on to another publishing house, we were equally lucky to work with Alex Littlefield, who completely grasped and embraced what we wanted to do and, through his inspired and meticulous editing, helped us achieve our vision.

Sydney Redepenning, one of Mike's doctoral students and a CIDRAP researcher, undertook yeoman's work in researching and checking every reference and citation throughout this book and providing critical assistance in completing the manuscript. A valued member of the CIDRAP team, she also serves as coproducer of the *Osterholm Update.*

Special thanks to all those who shared their knowledge, wisdom, and expertise with us: John Barry, Seth Berkeley, Robert Breiman, John Brooks, Lisa Brosseau, Christopher Chadwick, Dawn Duncan, Ezekiel Emanuel, Jason Furman, Sandro Galea, Bruce Gellin, Julie Gerberding, Edward Gillen, Richard Hatchett, Penny Heaton, James Hodge, Peter Hotez, Neel Kashkari, Michael Klompas, Florian Krammer, Marc Lipsitch, Ken Martinez, Nikki McCullough, Michelle Mello, Kristine Moore, Kariuki Njenga, Stanley Perlman, Sandra Quinn, Anne Schuchat, Lone Simonsen, Andy Slavitt, Gavin Smith, Brad Spellberg, Eric Topol, and Susan Wolf.

Much appreciation to Stewart Simonson for his critical review of our manuscript and enlightened suggestions.

Three titans of public health mentioned in the book continue to inspire us: Dr. William "Bill" Foege and the late Drs. Donald Ainslie "D.A." Henderson and Alexander "Alex" Langmuir. Among their many accomplishments and contributions, D.A. and Bill partnered in eradicating smallpox, probably the greatest gift the public health field has given to the world, and Alex conceived, established, and headed up the CDC's Epidemic Intelligence Service of medical detectives. They are among the giants on whose shoulders we stand.

MIKE:

No one just writes a book during a three-year pandemic like the one we have just experienced. We have all spent great energy, effort, pain, and suffering just to survive this horrible experience. Family, work colleagues, and dear friends got me through the pandemic emotionally and physically so I could focus energy on *The Big One.* Fern Peterson, my loving life partner and emotional rock, provides more support than I can put into words. My family—including my and Fern's kids Erin Osterholm, Ryan Osterholm, Jack Peterson, Will Peterson, and Tripp Peterson and their partners, Chad Pratt, Monica Osterholm, Annie Kuster, and Rachel Becker, and six special grandchildren, Connor, Thomas, Henry, Ethan, Greta, and James—was my loving fortress against all the evil this virus brought into our lives. Peggy Johnston, Craig Barness, Dave and Pat Pratt, and David and Joy Roslien completed the family cocoon.

The CIDRAP team members were incredible partners in our response to COVID, teaching me both the science and social lessons I needed to know. In particular, Kristine Moore, Jill DeBoer, Eve Lackritz, Carlos Cruz, and James Wappes were the leadership team that worked tirelessly to make CIDRAP activities consequential, comprehensive, and helpful to the public throughout the pandemic. Our CIDRAP News team and the *Osterholm Update* podcast were unique, credible, and compassionate voices of factual pandemic information. CIDRAP members providing invaluable assistance include Cory Anderson, Meredith Arpey, Lauren Bigalke, Chris Dall, Elise Holmes, Anje Mehr, Leah Moat, Sarah Morris, Laurel O'Neil, Julie Ostrowsky, Maya Peters, Marnie Peterson, Lisa Schnirring, Stephanie Soucheray, Clare Stoddart, Maria Sundaram, Angela Ulrich, Jamie Umber, Mary Van Beusekom, and Natalie Vestin.

Beginning in March 2020, I have had the incredible opportunity and privilege to host nearly two hundred episodes of the *Osterholm Update* podcast. These podcasts are our effort at CIDRAP to provide to the public, as well as colleagues, straight talk about the pandemic and other emerging infectious disease issues. The group of faithful listeners has numbered more than one hundred thousand. All have become part of a very special family for me. Thank you for all of your kind support and inspiring feedback.

Several valued colleagues—Ruth Berkelman, Bruce Gellin, Margaret "Peggy" Hamburg, Penny Heaton, Peter Hotez, and Eric Topol—joined me in the "Party Planning Group," where we held weekly confidential calls addressing the latest science, policy, and unofficial personal therapy issues of the day.

The Minnesota Department of Health, my professional home from 1975 to 1999, was a national leader in the public health response to the Covid pandemic. Commissioner Jan Malcolm and her team, led by Ruth Lynfield and Kristen Ehresmann, provided critical leadership to the state and the nation. Their including me in many of their response activities provided me with a remarkable learning experience. I'm so grateful for our collaboration.

Paul and Margot Andress, Amy Spomer and Sal Yelkin, Larry Montan and Jerry Johnson, John and Ann Norton, and Shawn Williams and Randy Knutson, provided immeasurable support, care, and needed good humor. Phillip Peterson, David Williams, Jay Coonan, Don Birkeland, and Stephen Thomas each shared moments of strength and kindness when I needed it most. And finally, Sarah Marie Taylor brought rainbows to my pandemic experience.

MARK:

This book began with a phone conversation I had during Covid with my close friend and fellow author Laurence "Larry" Leamer, during which I commented, "As bad as this is, it's not nearly the Big One." "That's your next book!" Larry responded. "And that's the title." A quick call to Mike secured his enthusiastic agreement; encouragement from Frank Weimann and a single-page outline and Zoom call with Tracy Behar set it all in motion.

Larry is a key member of my inner circle of writers who indefatigably support one another and glory in one another's successes, including Bob Bates, Marty Bell, Eric Dezenhall, Jim Grady, Nancy Lubin, Dan Moldea, Sally Rosenthal, Gus Russo, Joel Swerdlow, and Bennett Tramer.

I can always count on the knowledge, experience, and wise counsel of my two medical brothers, Drs. Robert and Jonathan Olshaker, and Robert's wife, Dr. Jacqueline Laurin. Their careers and care for their patients have been a living tribute to my late father, Dr. Bennett Olshaker.

Throughout my career, I have benefited from the collaboration of my film-producing partner Larry Klein. As always, Larry's influence is reflected throughout this book.

During the long months and years of the SARS-CoV-2 pandemic, I was sustained and nurtured by our small "Covid family": Neil and Damon Alpert, Cippy Klionsky and Jonny Slemrod, and Morgan Ortagus and Jonathan Weinberger. You six, together with a joyous new arrival from each couple during this time, "helped me make it through the night," and I am eternally grateful.

Finally, I would like to reprise my closing comment from our last book:

My wife, Carolyn, has been not only my partner through all things but an enthusiastic fellow traveler through all the adventures, as well as attorney, manager, counselor, and inspirer. I love you more than anything and couldn't have done it without you.

APPENDIX:
THOUGHT EXPERIMENT
CAST OF CHARACTERS

NAME	TITLE/ROLE
Adebayo, Kolawole (Dr.)	WHO director-general
Allen, Herb	Executive producer, *The Evening Roundup*
Amayana, Benjamin (Dr.)	Physician in Eastleigh suburb of Nairobi, Kenya
Andrews, Robert (Dr.)	Physician; expert who frequently appears in media, affiliated with the University of Michigan; "Bad News Bob"
Ashworth, Curt & Family	College student; volunteer with international outreach program in Kenya Caleb Ashworth, father of Curt Helen Ashworth, mother of Curt
Babcock, Paul	Journalist
Ballard, Jonathan (Dr.)	Director of the CDC's National Center for Emerging and Zoonotic Infectious Diseases
Banks, Tom	Journalist, Reuters

NAME	TITLE/ROLE
Bennett, Eugene	Former Speaker of the House
Borges, Alejandro "Andro"	Director of the White House Office of Science and Technology Policy
Brenner, Kate	Journalist, reporter, *The Guardian*
Bryce-Colvin, Lisa	Chief booker, *The Evening Roundup*
Caldwell, Brian (Dr.)	Director, National Institutes of Health
Cipora, Hannah	Host, *The Evening Roundup*
Davies, Jeremy (Dr.)	Executive director of WHO's Health Emergencies Programme in Geneva
Eggleston, Arthur	Secretary of labor
Engler, Ronald	First head of SARS-3 coordination
Goldfield, Tamara (Dr.)	CDC director
Gonzales, Nadia	CNN International reporter
Goodwin, Jonathan	Host, *Meet the Press*
Kimani, Adamu (Dr.)	Director of Kenya's National Virology Reference Lab
Mahad, Fawzia Noor & Family	Pregnant wife of Nadifa Nadifa Bashiir Ahmed, murdered husband of Fawzia
Malone, Caitlin (Dr.)	Department of Health and Human Services assistant secretary for preparedness and response (ASPR)
Marsh, Kara	National security advisor
McNeil, Damon	News director, *The Evening Roundup*
Minekal, Peter (Dr.) & Family	Vascular surgeon; bestselling author on nutrition and supplements; talking head Michaela Minekal, daughter of Peter Vivian Minekal, wife of Peter

NAME	TITLE/ROLE
Morgan, Deann	White House press secretary
Navarre, Etienne	French aid worker for an NGO; volunteer at Hagadera Refugee Camp
Ndembi, Joseph (Dr.)	Chief epidemiologist, Eastern Africa Regional Coordination Centre Headquarters, Kenyatta National Hospital, Nairobi, Kenya
O'Malley, Arnold	Education secretary
Onyango, Daniel (Dr.)	Chief health manager, International Rescue Committee, Hagadera Refugee Camp
Osman, Warsame Amir & Family	Somali farmer; husband of Yasmiin Cabdi Warsame Amir, middle son of Warsame and Yasmiin Yasmiin Said Ibrahim, wife of Warsame
Richman, Paul (Dr.)	Head of the Office of Pandemic Preparedness and Response Policy (OPPR)
Russell, Caleb M., Brigadier General	Leader of SARS-3 coordination after Ronald Engler's dismissal
Sadlock, Peter (Dr.)	Director of National Institute of Allergy and Infectious Diseases
Salat, Abdirahim Ali & Family	Somali IT graduate who travels to Minneapolis–St. Paul for work; husband of Bishaaro Abshir Ali Salat, brother of Abdirahim; husband of Caaliyah Bishaaro Abdullahi Adan, wife of Abdirahim Caaliyah Hamza Khalid, wife of Abshir; sister-in-law of Abdirahim

NAME	TITLE/ROLE
Scott, Erica (Dr.)	Hospitalist, Marin General Hospital
Shamshi, Jamilah	Somali community health worker
St. Clair, Jamie	Journalist, Associated Press
Stern, Gilbert	White House chief of staff
Sulbarry, Anne (Dr.)	Health and Human Services secretary; well-known virologist and vaccinologist who had her own lab at the National Institutes of Health; former visiting professor at the Johns Hopkins University School of Medicine and the Hopkins Bloomberg School of Public Health
Thomas, Erin (Dr.)	Attending physician in charge of shift at ER in Minneapolis–St. Paul; sees Abdirahim Salat and his brother, Abshir
Trowbridge, Jacinda (Dr.)	Principal deputy director, CDC
Weinberger, Ross	Homeland Security secretary
Wenda, Taramin	Indonesian businessman who travels throughout Kenya before returning home to Jakarta
Winters, Edward (Dr.)	Chairman of the Council of Economic Advisers
Yussef, Axlam Omar & Family	Somali farmer; husband of Zahi Hani Axlam Omar, one-year-old daughter of Axlam and Zahi; first child to die of new illness in Hagadera Mohamed Axlam Omar, three-year-old son of Axlam and Zahi Zahi Jama Hibaq, wife of Axlam

NOTES

PROLOGUE

1   "ProMED: Pneumonia Cases of Unknown Cause in Wuhan, China," ProMED, January 5, 2020, https://cepi.net/promed-pneumonia-cases -unknown-cause-wuhan-china.

2   Isaac I. Bogoch, Alexander Watts, Andrea Thomas-Bachli, Carmen Huber, Moritz U.G. Kraemer, and Kamran Khan, "Pneumonia of Unknown Aetiology in Wuhan, China: Potential for International Spread via Commercial Air Travel," *Journal of Travel Medicine* 27, no. 2 (2020): taaa008.

3   Meredith S. Shiels, Anika T. Haque, Amy Berrington de González, and Neal D. Freedman, "Leading Causes of Death in the US during the COVID-19 Pandemic, March 2020 to October 2021," *JAMA Internal Medicine* 182, no. 8 (2022).

4   Dimple D. Rajgor, Meng Har Lee, Sophia Archuleta, Natasha Bagdasarian, and Swee Chye Quek, "The Many Estimates of the COVID-19 Case Fatality Rate," *Lancet Infectious Diseases* 20, no. 7 (2020): 776–77, https://www .thelancet.com/journals/laninf/article/PIIS1473-3099(20)30244-9 /fulltext.

5   T.M. Abdelghany, Magdah Ganash, Marwah M. Bakri, Husam Qanash, Aisha M.H. Al-Rajhi, and Nadeem I. Elhussieny, "SARS-CoV-2, the Other Face to SARS-CoV and MERS-CoV: Future Predictions," *Biomedical Journal* 44, no. 1 (2021): 86–93.

6   "Middle East Respiratory Syndrome Coronavirus (MERS-CoV)," World Health Organization (WHO), https://www.who.int/news-room/fact -sheets/detail/middle-east-respiratory-syndrome-coronavirus-(mers-cov).

7   Dan Jones, "More People Died in the 1918 Flu Pandemic than in WWI," April 19, 2023, https://www.history.com/news/spanish-flu.

8   Max Roser, "The Spanish Flu: The Global Impact of the Largest Influenza

Pandemic in History," Our World in Data, https://ourworldindata.org /spanish-flu-largest-influenza-pandemic-in-history.

9   Jones, "More People Died in the 1918 Flu Pandemic."

10 John M. Barry, "The Site of Origin of the 1918 Influenza Pandemic and Its Public Health Implications," *Journal of Translational Medicine* 2, no. 3 (2004).

11 "Data on the Size of the HIV Epidemic," WHO World Health Statistics— HIV, https://www.who.int/data/gho/data/themes/topics/topic-details /GHO/data-on-the-size-of-the-hiv-aids-epidemic.

12 E. Thomas Ewing, "Measuring Mortality in the Pandemics of 1918–19 and 2020–21," *Health Affairs Forefront*, April 1, 2021.

13 Institute of Medicine of the National Academies, *The Threat of Pandemic Influenza: Are We Ready? Workshop Summary*, ed. Stacey L. Knobler, Alison Mack, Adel Mahmoud, and Stanley M. Lemon (Washington, DC: National Academies Press, 2005), chap. 1, "The Story of Influenza," 69.

14 "Ebola Virus Disease," Pan American Health Organization/WHO, https: //www.paho.org/en/topics/ebola-virus-disease; "Marburg Virus Disease," WHO, https://www.who.int/news-room/fact-sheets/detail/marburg -virus-disease.

15 Henry Kyobe Bosa, Neema Kamara, Merawi Aragaw, et al., "The West Africa Ebola Virus Disease Outbreak: 10 Years On," *Lancet Global Health* 12, no. 7 (2024): E1081–83.

16 Michael T. Osterholm, Kristine A. Moore, Nicholas S. Kelley, et al., "Transmission of Ebola Viruses: What We Know and What We Do Not Know," *mBio* 6, no. 2 (2015).

17 Miguel Ángel Muñoz-Alía, Rebecca A. Nace, Lianwen Zhang, and Stephen J. Russell, "Serotypic Evolution of Measles Virus Is Constrained by Multiple Co-dominant B Cell Epitopes on Its Surface Glycoproteins," *Cell Reports Medicine* 2, no. 4 (2021): 100225, https://www.cell.com/cell -reports-medicine/fulltext/S2666-3791(21)00041-0.

18 Raysa Rosario-Acevedo, Sergei S. Biryukov, Joel A. Bozue, and Christopher K. Cote, "Plague Prevention and Therapy: Perspectives on Current and Future Strategies," *Biomedicines* 9, no. 10 (2021): 1421.

19 Christina Newman, Thomas C. Friedrich, and David H. O'Connor, "Macaque Monkeys in Zika Virus Research: 1947–Present," *Current Opinion in Virology* 25 (2017): 34–40.

20 Beatriz Parra, Jairo Lizarazo, Jorge A. Jiménez-Arango, et al., "Guillain-Barré Syndrome Associated with Zika Virus Infection in Colombia," *New England Journal of Medicine (NEJM)* 375, no. 16 (2016): 1513–23.

21 Bruna Luiza de Amorin Vilharba, Mellina Yamamura, Micael Viana de Azevedo, Wagner de Souza Fernandes, Cláudia Du Bocage Santos-Pinto, and Everton Falcão de Oliveira, "Disease Burden of Congenital Zika

Virus Syndrome in Brazil and Its Association with Socioeconomic Data," *Scientific Reports* 13, no. 11882 (2023).

CHAPTER ONE

1   Yinon M. Bar-On, Avi Flamholz, Rob Phillips, and Ron Milo, "SARS-CoV-2 (COVID-19) by the Numbers," *eLife* 9 (2020): e57309.
2   James D. Cherry and Paul Krogstad, "SARS: The First Pandemic of the 21st Century," *Pediatric Research* 56, no. 1 (2004): 1–5.
3   "Flu Outbreaks Reminder of Pandemic Threat," Forum on Microbial Threats, World Bank Group, March 5, 2013, https://www.worldbank.org /en/news/feature/2013/03/05/flu-outbreaks-reminder-of-pandemic -threat.
4   Institute of Medicine, *Learning from SARS: Preparing for the Next Disease Outbreak — Workshop Summary,* ed. Stacey Knobler, Adel Mahmoud, Stanley Lemon, Alison Mack, Laura Sivitz, and Katherine Oberholtzer (Washington, DC: National Academies Press, 2004), 6–9.
5   Michael T. Osterholm and Mark Olshaker, *Deadliest Enemy: Our War against Killer Germs* (New York: Little, Brown Spark, 2020), 173–74.
6   "Rabies," WHO, https://www.who.int/health-topics/rabies#tab=tab_1.
7   "Frequently Asked Questions about SARS," CDC, May 3, 2005 (archived), accessed October 31, 2024, https://archive.cdc.gov/www_cdc_gov/sars /about/faq.html; "About Middle East Respiratory Syndrome (MERS)," CDC, May 28, 2024, accessed October 31, 2024, https://www.cdc.gov/mers /about/index.html.

CHAPTER TWO

1   Michael A.E. Ramsay, "John Snow, MD: Anaesthetist to the Queen of England and Pioneer Epidemiologist," *Baylor University Medical Center Proceedings* 19, no. 1 (2006): 24–28.
2   Charles Creighton, *A History of Epidemics in Britain,* vol. 2, *From the Extinction of Plague to the Present Time* (Cambridge: Cambridge University Press, 1894), 854.
3   K.R. Ehresmann, C.W. Hedberg, M.B. Grimm, C.A. Norton, K.L. MacDonald, and M.T. Osterholm, "An Outbreak of Measles at an International Sporting Event with Airborne Transmission in a Domed Stadium," *Journal of Infectious Diseases* 171, no. 3 (1995): 679–83.
4   Derek Thompson, "Hygiene Theater Is a Huge Waste of Time," *The Atlantic,* July 27, 2020.
5   SARS Commission, "Executive Summary," *Spring of Fear,* vol. 1, Commission to Investigate the Introduction and Spread of SARS in Ontario, The Honourable Mr. Justice Archie Campbell (Commissioner), December 11, 2006.

6   "Middle East Respiratory Syndrome Coronavirus (MERS-CoV)," WHO, August 5, 2022, https://www.who.int/health-topics/middle-east -respiratory-syndrome-coronavirus-mers#tab=tab_1.

7   WHO (@WHO), "FACT: #COVID19 is NOT airborne," Twitter, March 28, 2020, 2:44 p.m., https://x.com/WHO/status/1243972193169616898?lang =en.

8   Lidia Morawska and Donald K. Milton, "It Is Time to Address Airborne Transmission of Coronavirus Disease 2019 (COVID-19)," *Clinical Infectious Diseases* 71, no. 9 (2020): 2311–13.

9   Doctors in Unite, "Where's the Evidence Dr. Nabarro?," January 26, 2023, https://doctorsinunite.com/2023/01/26/wheres-the-evidence -dr-nabarro/.

10 Kai Kupferschmidt, "WHO's Departing Chief Scientist Regrets Errors in Debate over Whether SARS-CoV-2 Spreads through Air," *ScienceInsider*, November 23, 2022.

11 Lisa Brosseau, interview with the authors, October 18, 2022.

12 Chad J. Roy and Donald K. Milton, "Airborne Transmission of Communicable Infection—the Elusive Pathway," *NEJM* 350, no. 17 (2004): 1710–12.

13 Adams's tweet has since been deleted, but it was covered in multiple media sources, such as the *USA Today* example cited here: John Bacon, "'Seriously People—STOP BUYING MASKS!' Surgeon General Says They Won't Protect from Coronavirus," *USA Today*, March 2, 2020.

14 National Academies of Sciences, Engineering, and Medicine, *Rapid Expert Consultation on the Possibility of Bioaerosol Spread of SARS-CoV-2 for the COVID-19 Pandemic* (Washington, DC: National Academies Press, 2020), https://doi.org/10.17226/25769.

15 "Recommendation Regarding the Use of Cloth Face Coverings, Especially in Areas of Significant Community-Based Transmission," CDC, April 3, 2020, accessed October 24, 2024, https://stacks.cdc.gov/view/cdc /86440.

16 "Understanding the Difference," CDC/NIOSH, https://www.cdc.gov /niosh/npptl/pdfs/UnderstandDifferenceInfographic-508.pdf.

17 Brosseau interview.

18 Marina Fang, "CDC Director: Masks Are 'the Most Important, Powerful Public Health Tool We Have,'" *Huffington Post*, September 16, 2020; updated September 17, 2020.

19 Lisa M. Brosseau, Angela Ulrich, Kevin Escandón, Cory Anderson, and Michael T. Osterholm, "Commentary: What Can Masks Do? Part 1: The Science Behind COVID-19 Protection," CIDRAP, October 14, 2021, https: //www.cidrap.umn.edu/covid-19/commentary-what-can-masks-do-part -1-science-behind-covid-19-protection.

20 William C. Hinds and Yifang Zhu, *Aerosol Technology: Properties, Behavior, and Measurement of Airborne Particles,* 3rd ed. (Hoboken, NJ: John Wiley & Sons, 2022); "NIOSH-Approved Particulate Filtering Facepiece Respirators," CDC, July 12, 2024, accessed October 25, 2024, https://www.cdc.gov/niosh/npptl/topics/respirators/disp_part/default.html.

21 Dee DePass, "3M's Case Study in Mobilization," *Minnesota Star Tribune,* April 26, 2020.

22 "NIOSH-Approved N95 Particulate Filtering Facepiece Respirators," CDC, https://www.cdc.gov/niosh/npptl/topics/respirators/disp_part/n95list1.html.

23 "Strategies for Optimizing the Supply of N95 Respirators," CDC, September 16, 2021, accessed October 27, 2024, https://archive.cdc.gov/www_cdc_gov/coronavirus/2019-ncov/hcp/respirators-strategy/index.html.

24 Michael S. Bergman, Dennis J. Viscusi, Ziqing Zhuang, Andrew J. Palmiero, Jeffrey B. Powell, and Ronald E. Shaffer, "Impact of Multiple Consecutive Donnings on Filtering Facepiece Respirator Fit," *American Journal of Infection Control* 40, no. 4 (2012): 375–80.

25 Unpublished data from CIDRAP.

26 Andrew Jacobs, "A Glut of Chinese Masks Is Driving U.S. Companies Out of Business," *New York Times,* May 29, 2021.

27 "Revoked EUAs for Non-NIOSH-Approved Disposable Filtering Facepiece Respirators," US Food & Drug Administration, curated May 9, 2024, https://www.fda.gov/medical-devices/emergency-use-authorizations-medical-devices/revoked-euas-non-niosh-approved-disposable-filtering-facepiece-respirators.

28 *Getting To and Sustaining the Next Normal: A Roadmap for Living with COVID,* United States, 2022, web archive, https://www.loc.gov/item/lcwaN0039464/.

29 Brosseau interview.

30 Brosseau interview.

31 Brosseau interview.

32 Aden Tate, "How *The Cobra Event* Changed American Pandemic Response," *MIRA Safety* (blog), July 5, 2023, accessed November 1, 2024, https://www.mirasafety.com/blogs/news/how-the-cobra-event-changed-american-pandemic-response?srsltid=AfmBOop0_KPPih4MxQtx_DQjLLnEAIMHhIOVFKXZVPG_kDK2FbUSmkIO.

33 "Respirators Beyond Their Shelf Life—Considerations," 3M Technical Bulletin, revision 3, May 2020, https://multimedia.3m.com/mws/media/1807271O/respirators-beyond-their-shelf-life-considerations-technical-bulletin.pdf.

34 Denise M. Hinton, Chief Scientist, Food and Drug Administration, to Rochelle P. Walensky, Director, Centers for Disease Control and Prevention, July 12, 2021, https://www.fda.gov/media/135763/download.

35 National Academies of Sciences, Engineering, and Medicine, *Building Resilience into the Nation's Medical Product Supply Chains* (Washington, DC: National Academies Press, 2022), 27, 70–73.

36 E.C. Riley, G. Murphy, and R.L. Riley, "Airborne Spread of Measles in a Suburban Elementary School," *American Journal of Epidemiology* 107, no. 5 (1978): 421–32.

37 Joseph G. Allen and Andrew M. Ibrahim, "Indoor Air Changes and Potential Implications for SARS-CoV-2 Transmission," *JAMA* 325, no. 20 (2021): 2112–13.

38 "ASHRAE Approves Groundbreaking Standard to Reduce the Risk of Disease Transmission in Indoor Spaces," ASHRAE press release, June 24, 2023, https://www.ashrae.org/about/news/2023/ashrae-approves -groundbreaking-standard-to-reduce-the-risk-of-disease-transmission -in-indoor-spaces.

39 *2018 Commercial Buildings Energy Consumption Survey Final Results,* US Energy Information Administration (EIA), https://www.eia.gov /consumption/commercial/data/2018/bc/pdf/b1.pdf.

40 *2018 Manufacturing Energy Consumption Survey,* EIA, https://www.eia.gov /consumption/manufacturing/pdf/MECS%202018%20Results %20Flipbook.pdf.

41 *Department of Defense Base Structure Report—Fiscal Year 2017 Baseline,* https: //www.acq.osd.mil/eie/Downloads/BSI/Base%20Structure%20Report %20FY17.pdf.

42 *2018 Commercial Buildings Energy Consumption Survey Final Results.*

43 *2023 American Housing Survey,* US Census Bureau, https://www.census.gov /programs-surveys/ahs/data/interactive/ahstablecreator.html.

44 "BART: Safer than Ever," Bay Area Rapid Transit, July 18, 2022, accessed October 25, 2024, https://www.bart.gov/news/articles/2022/news 20220718-1.

45 "Enhanced Cleaning and Air Filtration Improvements for Covid-19," Washington Metropolitan Area Transit Authority, accessed November 25, 2022, https://www.wmata.com/service/covid19/covid19-cleaning.cfm.

46 Jordan Pascale, "Metro Receives Grant to Test Air Filtration and Install UV Lights in Rail Cars," *DCist,* January 28, 2021, accessed November 25, 2022, https://dcist.com/story/21/01/28/metro-air-quality-grant -coronavirus/.

47 Miguella Mark-Carew, Gloria Kang, Sanjana Pampati, Kenneth R. Mead, Stephen B. Martin Jr., and Lisa C. Barrios, "Ventilation Improvements among K–12 Public School Districts—United States, August–December 2022," *Morbidity and Mortality Weekly Report* (*MMWR*) 72, no. 14 (2003): 372–76.

48 Mark-Carew et al., "Ventilation Improvements."

49 Hinds and Zhu, *Aerosol Technology;* Environmental Protection Agency,

"What Is a MERV Rating?," March 5, 2024, https://www.epa.gov/indoor
-air-quality-iaq/what-merv-rating.

50 Mark-Carew et al., "Ventilation Improvements," 373–74.

51 T.Q. Corrêa, K.C. Blanco, J.D. Vollet-Filho, V.S. Morais, W.R. Trevelin, S.
Pratavieira, and V.S. Bagnato, "Efficiency of an Air Circulation
Decontamination Device for Micro-organisms Using Ultraviolet
Radiation," *Journal of Hospital Infection* 115 (2021): 32–43.

CHAPTER THREE

1   Janet A. Aker, "Gen. George Washington Ordered Smallpox Inoculations
for All Troops," MHS Communications, https://www.health.mil/News
/Articles/2021/08/16/Gen-George-Washington-Ordered-Smallpox
-Inoculations-for-All-Troops.

2   Dan Liebowitz, "Smallpox Vaccination: An Early Start of Modern
Medicine in America," *Journal of Community Hospital Internal Medicine
Perspective* 7, no. 1 (2017): 61–63.

3   "Smallpox, Inoculation, and the Revolutionary War," National Park
Service, https://www.nps.gov/articles/000/smallpox-inoculation
-revolutionary-war.htm.

4   Dave Roos, "How Crude Smallpox Inoculations Helped George
Washington Win the War," *History,* June 20, 2023, https://www.history
.com/news/smallpox-george-washington-revolutionary-war.

5   "Smallpox, Inoculation, and the Revolutionary War."

6   John Marshall Harlan and Supreme Court of the United States, US
Reports: *Jacobson v. Massachusetts,* 197 US 11 (1905), https://www.loc
.gov/item/usrep197011/.

7   National Academy of Sciences, *The Future of Public Health* (Washington,
DC: National Academy Press, 1988), chap. 3, "A History of the Public
Health System," 57.

8   National Academy of Sciences, *Future of Public Health,* 61.

9   National Academy of Sciences, *Future of Public Health,* 61, 63.

10 National Academy of Sciences, *Future of Public Health,* 60.

11 National Academy of Sciences, *Future of Public Health,* 66.

12 Carol R. Byerly, "The U.S. Military and the Influenza Pandemic of 1918–
1919," *Public Health Reports* 125, supp. 3 (2010): 82–91.

13 John Barry, interview with the authors, February 14, 2023.

14 George A. Soper, "The Influenza Pneumonia Pandemic in the American
Army Camps during September and October, 1918," *Science,* n.s., 48, no.
1245 (1918): 454.

15 *The Medical Department of the United States Army in the World War,* vol. 9,
Communicable and Other Diseases, prepared under the Direction of Maj. Gen.
M.W. Ireland, the Surgeon General, by Lieut. Col. Joseph F. Siler, M.C., US
Army (Washington, DC: US Government Printing Office, 1928),

"Measures Designed to Prevent the Entrance and Spread of Infection in a Command—Quarantine," 116, 123–24.

16   John M. Barry, "1918 Revisited: Lessons and Suggestions for Further Inquiry," in *The Threat of Pandemic Influenza: Are We Ready? Workshop Summary,* Institute of Medicine (US) Forum on Microbial Threats, ed. S.L. Knobler, A. Mack, A. Mahmoud, et al. (Washington, DC: National Academies Press, 2005).

17   Filio Marineli, Gregory Tsoucalas, Marianna Karamanou, and George Androutsos, "Mary Mallon (1869–1938) and the History of Typhoid Fever," *Annals of Gastroenterology* 26, no. 2 (2013): 132–34.

18   Marineli et al., "Mary Mallon."

19   George A. Soper, "The Curious Career of Typhoid Mary," *Bulletin of the New York Academy of Medicine* 15, no. 10 (1939): 712.

20   Janet Brooks, "The Sad and Tragic Life of Typhoid Mary," *Canadian Medical Association Journal* 154, no. 6 (1996): 916.

21   Brooks, "Sad and Tragic Life," 915–16.

22   George A. Soper, "The Work of a Chronic Typhoid Germ Distributor," *JAMA* 48, no. 24 (1907): 2019–22.

23   Marineli et al., "Mary Mallon," 134.

24   Mark Honigsbaum, "The Art of Medicine: Revisiting the 1957 and 1968 Influenza Pandemics," *Lancet* 395, no. 10240 (2020): 1824–26; Ravi Jhaveri, "Echoes of 2009 H1N1 Influenza Pandemic in the COVID Pandemic," *Clinical Therapeutics* 42, no. 5 (2020): 736–40.

25   Jeffrey A. Tucker, "They Considered and Rejected Pandemic Closures in 1957," *Brownstone Journal,* August 2, 2021, https://brownstone.org/articles /they-considered-and-rejected-pandemic-closures-in-1957/.

26   Dan E. Beauchamp, "Community: The Neglected Tradition of Public Health," *The Hastings Center Report* 15, no. 6 (1985): 28–36.

27   Michelle M. Mello, interview with the authors, January 13, 2023.

28   "Disease Outbreak News: Pandemic (H1N1) 2009—Update 58," WHO, July 6, 2009, https://www.who.int/emergencies/disease-outbreak-news /item/2009_07_06-en.

29   "2009 H1N1 Pandemic (H1N1 pdm09 virus)," CDC, June 11, 2019, https: //archive.cdc.gov/#/details?url=https://www.cdc.gov/flu/pandemic -resources/2009-h1n1-pandemic.html; Nobuyuki Horita and Takeshi Fukumoto, "Global Case Fatality Rate from COVID-19 Has Decreased by 96.8% during 2.5 Years of the Pandemic," *Journal of Medical Virology* 95, no. 1 (2023): e28231.

30   Noreen Qualls, Alexandra Levitt, Neha Kanade, Narue Wright-Jegede, Stephanie Dopson, Matthew Biggerstaff, Carrie Reed, and Amra Uzicanin, "Community Mitigation Guidelines to Prevent Pandemic Influenza—United States, 2017," *MMWR* 66, no. 1 (2017), https://www .cdc.gov/mmwr/volumes/66/rr/pdfs/rr6601.pdf.

31 Maureen Rubin, "Supreme Court Says Churches Can't Be Forced to Close during COVID-19," *Law Commentary,* February 11, 2021.

32 David Choi, "Los Angeles County Orders Ambulance Crews to 'Conserve Oxygen' as It Deals with a 905% Increase in COVID-19 Cases," *Business Insider,* January 5, 2021; Jaclyn Diaz, "LA County Paramedics Told Not to Transport Some Patients with Low Chance of Survival," NPR News on WBUR, January 5, 2021, accessed September 16, 2023, https://www.wbur .org/npr/953444637/l-a-paramedics-told-not-to-transport-some-patients -with-low-chance-of-survival.

33 "Updated WHO Recommendations for International Traffic in Relation to COVID-19 Outbreak," WHO, February 29, 2020, https://www.who.int /news-room/articles-detail/updated-who-recommendations-for -international-traffic-in-relation-to-covid-19-outbreak.

34 WHO, "Statement on the Second Meeting of the International Health Regulations (2005) Emergency Committee Regarding the Outbreak of Novel Coronavirus (2019-nCoV)," January 30, 2020, https://www.who.int /news/item/30-01-2020-statement-on-the-second-meeting-of-the -international-health-regulations-(2005)-emergency-committee -regarding-the-outbreak-of-novel-coronavirus-(2019-ncov).

35 "Updated WHO Recommendations for International Traffic in Relation to COVID-19 Outbreak."

36 Mary A. Shiraef, Paul Friesen, Lukas Feddern, Mark A. Weiss, and COBAP Team, "Did Border Closures Slow SARS-CoV-2?," *Scientific Reports* 12, no. 1 (2022): 1709.

37 "DHS Issues Supplemental Instructions for Inbound Flights with Individuals Who Have Been in China," Department of Homeland Security, February 2, 2020, https://www.dhs.gov/archive/news/2020/02 /02/dhs-issues-supplemental-instructions-inbound-flights-individuals -who-have-been.

38 Corinne N. Thompson, Jennifer Baumgartner, Carolina Pichardo, et al., "COVID-19 Outbreak — New York City, February 29–June 1, 2020," *Morbidity and Mortality Weekly Report* 69, no. 46 (2020): 1725–29.

39 Jing Yang, Juan Li, Shengjie Lai, et al., "Uncovering Two Phases of Early Intercontinental COVID-19 Transmission Dynamics," *Journal of Travel Medicine* 27, no. 8 (2020); Jeff Prince and Daniel Simon, "Travelers Coming from Italy May Have Driven First U.S. COVID-19 Wave More than Those from China, Study Suggests," *The Conversation,* January 29, 2021.

40 Yang et al., "Uncovering Two Phases."

41 Dorey Scheimer and Meghna Chakrabarti, "Why the U.S. Needs a Reckoning on Lockdowns Before the Next Pandemic," WBUR, *On Point,* June 11, 2024, https://www.wbur.org/onpoint/2024/06/11/pandemic -covid-lockdown-public-health-physician.

42   Peter Bergen, "The Infectious Disease Expert Who Warned Us 800,000
      Americans Would Die of Covid-19," CNN, December 11, 2021, https:
      //www.cnn.com/2021/12/09/opinions/infectious-disease-expert
      -warned-covid-19-deaths-bergen/index.html.

43   "COVID Data Tracker: Trends in United States COVID-19 Deaths,
      Emergency Department (ED) Visits, and Test Positivity by Geographic
      Area," CDC, https://covid.cdc.gov/covid-data-tracker/#trends
      _totaldeaths_select_00.

44   Daniel Wolfe and Daniel Dale, "'It's Going to Disappear': A Timeline of
      Trump's Claims that Covid-19 Will Vanish," CNN, October 31, 2020,
      accessed October 24, 2024, https://www.cnn.com/interactive/2020/10
      /politics/covid-disappearing-trump-comment-tracker/.

45   Toby Phillips, Yuxi Zhang, and Anna Petherick, "A Year of Living
      Distantly: Global Trends in the Use of Stay-at-Home Orders over the First
      12 Months of the COVID-19 Pandemic," *Interface Focus* 11, no. 6 (2021):
      20210041.

46   Emergency Executive Order 20-20, State of Minnesota Executive
      Department, March 25, 2020, https://mn.gov/governor/assets/3a.%20
      EO%2020-20%20FINAL%20SIGNED%20Filed_tcm1055-425020.pdf;
      Catharine Richert, "Essential Workers Say Employers Not Following
      Work-from-Home, Safety Precautions," MPR News, March 31, 2020, https:
      //www.mprnews.org/story/2020/03/31/essential-workers-say-employers
      -not-following-work-from-home-safety-precautions.

47   Max Boot, "New York Has Flattened Its Curve. The Sun Belt Has Been
      Flattened by It," *Washington Post,* July 27, 2020; Eric Toner, Vikramjit
      Mukherjee, Dan Hanfling, John Hick, Lee Daugherty Biddison, Amesh
      Adalja, Matthew Watson, and Laura Evans, *Crisis Standards of Care: Lessons
      from New York City Hospitals' COVID-19 Experience. A Meeting Report,* Johns
      Hopkins Center for Health Security, 2020.

48   Michael T. Osterholm and Mark Olshaker, "Facing Covid-19 Reality: A
      National Lockdown Is No Cure," *Washington Post,* March 21, 2020.

49   Michael T. Osterholm and Neel Kashkari, "Here's How to Crush the Virus
      Until Vaccines Arrive," *New York Times,* August 7, 2020.

50   Meagan C. Fitzpatrick, Seyed M. Moghadas, Abhishek Pandey, and Alison
      P. Galvani, "Two Years of U.S. COVID-19 Vaccines Have Prevented
      Millions of Hospitalizations and Deaths," *The Commonwealth Fund* (blog),
      December 13, 2022, https://www.commonwealthfund.org/blog/2022
      /two-years-covid-vaccines-prevented-millions-deaths-hospitalizations.

51   Emily Head and Sabine L. van Elsland, "Vaccinations May Have
      Prevented Almost 20 Million COVID-19 Deaths Worldwide," *Imperial
      News,* Imperial College London, June 24, 2022, https://www.imperial.ac
      .uk/news/237591/vaccinations-have-prevented-almost-20-million/#.

52 Han Sun, Yu Zhang, Ge Wang, Wen Yang, and Yingjie Xu, "mRNA-Based Therapeutics in Cancer Treatment," *Pharmaceutics* 15, no. 2 (2023): 622.

53 "Decades in the Making: mRNA COVID-19 Vaccines," NIAID: National Institute of Allergy and Infectious Diseases, April 4, 2023, https://www.niaid.nih.gov/diseases-conditions/decades-making-mrna-covid-19-vaccines.

54 Martin Alberer, Ulrike Gnad Vogt, Henoch Sangjoon Hong, et al., "Safety and Immunogenicity of a mRNA Rabies Vaccine in Healthy Adults: An Open-Label, Non-randomised, Prospective, First-in-Human Phase 1 Clinical Trial," *Lancet* 390, no. 10101 (2017): 1511–20.

55 "Gene-Based Zika Vaccine Is Safe and Immunogenic in Healthy Adults," National Institute of Allergy and Infectious Diseases, news release, December 4, 2017, https://wayback.archive-it.org/7880/20230226125329/https:/www.niaid.nih.gov/news-events/gene-based-zika-vaccine-safe-and-immunogenic-healthy-adults.

56 Fernando P. Polack, Stephen J. Thomas, Nicholas Kitchin, et al., "Safety and Efficacy of the BNT162b2 mRNA Covid-19 Vaccine," *NEJM* 383, no. 27 (2020): 2603–15.

57 "Moderna Announces FDA Authorization of Moderna COVID-19 Vaccine in U.S.," Moderna, news release, December 19, 2020, https://s29.q4cdn.com/435878511/files/doc_news/2020/12/18/moderna-announces-fda-authorization-moderna-covid-19-vaccine-us.pdf.

58 Henry S. Sacks, "The Single-Dose J&J Vaccine Had 67% Efficacy against Moderate to Severe-Critical COVID-19 at ≥14 d," *Annals of Internal Medicine* 174, no. 7 (2021): JC75.

59 J. Sadoff, G. Gray, A. Vandebosch, et al., "Safety and Efficacy of Single-Dose Ad26.COV2.S Vaccine against Covid-19," *NEJM* 384, no. 23 (2021): 2187–201.

60 Polack et al., "Safety and Efficacy of the BNT162b2 mRNA Covid-19 Vaccine"; L.R. Baden, H.M. El Sahly, B. Essink, et al., "Efficacy and Safety of the mRNA-1273 SARS-CoV-2 Vaccine," *NEJM* 384, no. 5 (2020).

61 Eriko Padron-Regalado, "Vaccines for SARS-CoV-2: Lessons from Other Coronavirus Strains," *Infectious Diseases and Therapy* 9 (2020): 255–74; Hannah Chung, Michael A. Campitelli, Sarah A. Buchan, et al., "Measuring Waning Protection from Seasonal Influenza Vaccination during Nine Influenza Seasons, Ontario, Canada, 2010/11 to 2018/19," *Eurosurveillance* 29, no. 8 (2024): pii=2300239.

62 Polack et al., "Safety and Efficacy of the BNT162b2 mRNA Covid-19 Vaccine"; Baden et al., "Efficacy and Safety of the mRNA-1273 SARS-CoV-2 Vaccine."

63 "COVID-19 Vaccine Breakthrough Infections Reported to CDC—United States, January 1–April 30, 2021," *MMWR* 70 (2021): 792–93.

64   Daniel R. Feikin, Melissa M. Higdon, Laith J. Abu-Raddad, et al.,
      "Duration of Effectiveness of Vaccines against SARS-CoV-2 Infection and
      COVID-19 Disease: Results of a Systematic Review and Meta-regression,"
      Lancet 399, no. 10328 (2022): 924–44; Francesco Menegale, Mattia
      Manica, Agnese Zardini, Giorgio Guzzetta, Valentina Marziano, Valeria
      d'Andrea, Filippo Trentini, Marco Ajelli, Piero Poletti, and Stefano
      Merler, "Evaluation of Waning of SARS-CoV-2 Vaccine–Induced
      Immunity: A Systematic Review and Meta-analysis," *JAMA Network Open* 6,
      no. 5 (2023): e2310650.

65   Dan-Yu Lin, Yangjianchen Xu, Yu Gu, Donglin Zeng, Shadia K. Sunny,
      and Zack Moore, "Durability of Bivalent Boosters against Omicron
      Subvariants," *NEJM* 338 (2023): 1818–20.

66   Ruth Link-Gelles, Zachary A. Weber, Sarah E. Reese, et al., "Estimates of
      Bivalent mRNA Vaccine Durability in Preventing COVID-19–Associated
      Hospitalization and Critical Illness among Adults with and without
      Immunocompromising Conditions—VISION Network, September 2022–
      April 2023," *MMWR* 72, no. 21 (2023): 579–88.

67   Gillian K. SteelFisher, Mary G. Findling, Hannah L. Caporello, and
      Rebekah I. Stein, "Has COVID-19 Threatened Routine Childhood
      Vaccination? Insights from U.S. Public Opinion Polls," *Health Affairs
      Forefront,* June 6, 2023, https://www.healthaffairs.org/content/forefront
      /has-covid-19-threatened-routine-childhood-vaccination-insights-us
      -public-opinion-polls.

68   Andrea Ciaranello and Taison Bell, "Using Data and Modeling to
      Understand the Risks of In-Person Education," *JAMA Network Open* 4, no.
      3 (2021): e214619.

69   "COVID-NET Laboratory-Confirmed COVID-19 Hospitalizations," CDC,
      November 8, 2024, https://covid.cdc.gov/covid-data-tracker/#covidnet
      -hospitalization-network.

70   "Deaths by Select Demographic and Geographic Characteristics," CDC,
      September 27, 2023, https://www.cdc.gov/nchs/nvss/vsrr/covid_weekly
      /index.htm.

71   Yi-Ju Tseng, Karen L. Olson, Danielle Bloch, and Kenneth D. Mandl,
      "Smart Thermometer–Based Participatory Surveillance to Discern the
      Role of Children in Household Viral Transmission during the COVID-19
      Pandemic," *JAMA Network Open* 6, no. 6 (2023): e2316190, https://doi.org
      /10.1001/jamanetworkopen.2023.16190.

72   Clarissa Oeser, Heather Whitaker, Ezra Linley, et al., "Large Increases in
      SARS-CoV-2 Seropositivity in Children in England: Effects of the Delta
      Wave and Vaccination," *Journal of Infection* 84, no. 3 (2022): 427–30.

73   Danielle Allen and Ashish Jha, "We've Figured Out It's Safe to Have
      Schools Open. Keep Them That Way," *Washington Post,* November 19, 2020.

74 Emily Oster, "Schools Aren't Super-Spreaders: Fears from the Summer Appear to Have Been Overblown," *The Atlantic,* October 9, 2020.

75 "Child Vaccination Coverage and Parental Intent for Vaccination," CDC, November 6, 2024, https://www.cdc.gov/covidvaxview/weekly-dashboard /child-coverage-vaccination.html.

76 Kanecia O. Zimmerman, M. Alan Brookhart, Ibukunoluwa C. Kalu, et al. for the ABC Science Collaborative, "Community SARS-CoV-2 Surge and within-School Transmission," *Pediatrics* 148, no. 4 (2021): e2021052686; Terri Rebmann, Travis M. Loux, Lauren D. Arnold, Rachel Charney, Deborah Horton, and Ashley Gomel, "SARS-CoV-2 Transmission to Masked and Unmasked Close Contacts of University Students with COVID-19—St. Louis, Missouri, January–May 2021," *MMWR* 70, no. 36 (2021): 1245–48.

77 Lisa M Brosseau, Angela Ulrich, Kevin Escandón, Cory Anderson, and Michael T. Osterholm, "Commentary: What Can Masks Do? Part 2: What Makes for a Good Mask Study—and Why Most Fail," CIDRAP, October 15, 2021, https://www.cidrap.umn.edu/covid-19/commentary-what-can -masks-do-part-2-what-makes-good-mask-study-and-why-most-fail.

CHAPTER FOUR

1 Priya Joi, "10 Infectious Diseases That Could Be the Next Pandemic," *VaccinesWork,* May 7, 2020, https://www.gavi.org/vaccineswork/10 -infectious-diseases-could-be-next-pandemic.

2 M. Cristina Cassetti, Theodore C. Pierson, L. Jean Patterson, Karin Bok, Amanda J. DeRocco, Anne M. Deschamps, Barney S. Graham, Emily J. Erbelding, and Anthony S. Fauci, "Prototype Pathogen Approach for Vaccine and Monoclonal Antibody Development: A Critical Component of the NIAID Plan for Pandemic Preparedness," *Journal of Infectious Diseases* 227, no. 12 (2023): 1433–41.

3 Anne M. Deschamps, Amanda J. DeRocco, Karin Bok, and L. Jean Patterson, "Prototype Pathogens for Vaccine and Monoclonal Antibody Countermeasure Development: NIAID Workshop Process and Outcomes for Viral Families of Pandemic Potential," *Journal of Infectious Diseases* 228, supp. 6 (2023): S355–S358.

4 Stefan Riedel, "Edward Jenner and the History of Smallpox and Vaccination," *Proceedings (Baylor University Medical Center)* 18, no. 1 (2005): 21–25.

5 "James Phipps 1788–1853," Science Museum Group Collection, https: //collection.sciencemuseumgroup.org.uk/people/cp166975/james -phipps.

6 Riedel, "Edward Jenner."

7 Riedel, "Edward Jenner."

8   "Measles, Mumps, and Rubella (MMR) Vaccination: What Everyone Should Know," CDC, January 26, 2021, https://www.cdc.gov/vaccines/vpd/mmr/public/index.html.

9   Miguel Ángel Muñoz-Alía, Rebecca A. Nace, Lianwen Zhang, and Stephen J. Russell, "Serotypic Evolution of Measles Virus Is Constrained by Multiple Co-dominant B Cell Epitopes on Its Surface Glycoproteins," *Cell Reports Medicine* 2, no. 4 (2021): 100225, https://www.cell.com/cell-reports-medicine/fulltext/S2666-3791(21)00041-0.

10 Erica E. Zeno, Francisco Nogareda, Annette Regan, et al., "Interim Effectiveness Estimates of 2024 Southern Hemisphere Influenza Vaccines in Preventing Influenza-Associated Hospitalization—REVELAC-i Network, Five South American Countries, March–July 2024," *MMWR* 73, no. 39 (2024): 861–68.

11 Lakshmi Panagiotakopoulos, Danielle L. Moulia, Monica Godfrey, et al., "Use of COVID-19 Vaccines for Persons Aged ≥6 Months: Recommendations of the Advisory Committee on Immunization Practices—United States, 2024–2025," *MMWR* 73, no. 37 (2024): 819–24.

12 "Coronavirus (COVID-19) Deaths," *Our World in Data,* October 13, 2024, https://ourworldindata.org/covid-deaths.

13 Børge Brende, Jeremy Farrar, Diane Gashumba, Carlos Moedas, Trevor Mundel, Yasuhisa Shiozaki, Harsh Vardhan, Johanna Wanka, and John-Arne Røttingen, "CEPI—a New Global R&D Organisation for Epidemic Preparedness and Response," *Lancet* 389, no. 10066 (2017): 233–35.

14 "Why We Exist," CEPI, https://cepi.net/why-we-exist.

15 "CEPI Investors," CEPI, https://cepi.net/investors.

16 "What We Do," CEPI, https://cepi.net/what-we-do.

17 "What We Do."

18 "Priority Pathogens," CEPI, https://cepi.net/priority-pathogens.

19 Seth Berkley, MD, interview with the authors, May 9, 2023.

20 "COVID 19-Vaccine Tracker: 12 Vaccines Granted Emergency Use Listing (EUL) by WHO," December 2, 2022, https://covid19.trackvaccines.org/agency/who/.

21 Francesco Menegale, Mattia Manica, Agnese Zardini, Giorgio Guzzetta, Valentina Marziano, Valeria d'Andrea, Filippo Trentini, Marco Ajelli, Piero Poletti, and Stefano Merler, "Evaluation of Waning of SARS-CoV-2 Vaccine–Induced Immunity: A Systematic Review and Meta-analysis," *JAMA Network Open* 6, no. 5 (2023): e2310650.

22 Florian Krammer, PhD, interview with the authors, May 15, 2023.

23 Krammer interview.

24 Bruce Gellin, MD, MPH, interview with the authors, May 31, 2023.

25 Richard Hatchett, MD, interview with the authors, May 25, 2023.

26 Krammer interview.

27 "Timeline on the Pandemic (H1N1) 2009," European Centre for Disease Prevention and Control, August 11, 2010, https://www.ecdc.europa.eu/en /seasonal-influenza/2009-influenza-h1n1-timeline.

28 Harvey V. Fineberg, "Pandemic Preparedness and Response—Lessons from the H1N1 Influenza of 2009," *NEJM* 370, no. 14 (2014): 1335–42.

29 Clare Stroud, Lori Nadig, and Bruce M. Altevogt, Institute of Medicine, *The 2009 H1N1 Influenza Vaccination Campaign: Summary of a Workshop Series* (Washington, DC: National Academies Press, 2010), https://www .ncbi.nlm.nih.gov/books/NBK54181/; "Update: Influenza Activity— United States, August 30–October 31, 2009," *MMWR* 58, no. 44 (2009): 1236–41.

30 *Report of the WHO Pandemic Influenza A(H1N1) Vaccine Deployment Initiative,* WHO, January 1, 2012, https://www.who.int/publications/i/item/report -of-the-who-pandemic-influenza-a(h1n1)-vaccine-deployment-initiative.

31 Michael T. Osterholm, Nicholas S. Kelley, Alfred Sommer, and Edward A. Belongia, "Efficacy and Effectiveness of Influenza Vaccines: A Systematic Review and Meta-analysis," *Lancet Infectious Diseases* 12, no. 1 (2012), epub October 25, 2011. Erratum in *Lancet Infectious Diseases* 12, no. 9 (2012): 655.

32 Marie R. Griffin, Arnold S. Monto, Edward A. Belongia, et al., for the US Flu-VE Network, "Effectiveness of Non-Adjuvanted Pandemic Influenza A Vaccines for Preventing Pandemic Influenza Acute Respiratory Illness Visits in 4 U.S. Communities," *PLOS One* 6, no. 8 (2011): e23085.

33 Osterholm et al., "Efficacy and Effectiveness of Influenza Vaccines."

34 Michael T. Osterholm and Mark Olshaker, "We're Not Ready for a Flu Pandemic," *New York Times,* January 8, 2018.

35 *Accelerating the Development of a Universal Influenza Vaccine* (Washington, DC: Aspen Institute and Sabin Vaccine Institute, 2019), 24, https://www .sabin.org/app/uploads/2022/05/Sabin-Aspen_Influenza_Report.pdf.

36 *Influenza Vaccines R&D Roadmap,* Center for Infectious Disease Research and Policy (CIDRAP), University of Minnesota, September 2021, February 2023, https://ivr.cidrap.umn.edu/roadmap.

37 Global Funders Consortium for Universal Influenza Vaccine Development, https://unifluvac.org/about/.

38 Search result using National Institutes of Health Research Portfolio Online Reporting Tools (RePORT) RePORTER, https://reporter.nih.gov /search/T3dRITMaY02U2SyMg8jGGA/projects?fy=2023;2022;20 21;2020;2019.

39 "Fact Sheet: HHS Details $5 Billion 'Project NextGen' Initiative to Stay Ahead of COVID-19," HHS Press Office, May 11, 2023, https://www.hhs .gov/about/news/2023/05/11/fact-sheet-hhs-details-5-billion-project -nextgen-initiative-stay-ahead-covid.html.

40 Benjamin Mueller, Noah Weiland, and Carl Zimmer, "U.S. Vaccine Program Now Flush with Cash, but Short on Key Details," *New York Times*, June 26, 2023.

41 Gellin interview.

42 Gellin interview.

43 Berkley interview.

44 Gellin interview.

45 Christopher Chadwick, MS, interview with the authors, June 15, 2023.

46 Chadwick interview.

47 "WHO and MPP Announce Names of 15 Manufactures [*sic*] to Receive Training from mRNA Technology Transfer Hub," Medicines Patent Pool (MPP) news release, April 19, 2022, https://medicinespatentpool.org/news-publications-post/who-and-mpp-announce-names-of-15-manufactures-to-receive-training-from-mrna-technology-transfer-hub.

48 Chadwick interview.

49 Erin Sparrow, James G. Wood, Christopher Chadwick, Anthony T. Newall, Siranda Torvaldsen, Ann Moen, and Guido Torelli, "Global Production Capacity of Seasonal and Pandemic Influenza Vaccines in 2019," *Vaccine* 39, no. 3 (2021): 512–20.

50 Fernando P. Polack, Stephen J. Thomas, Nicholas Kitchin, et al., "Safety and Efficacy of the BNT162b2 mRNA Covid-19 Vaccine," *NEJM* 383, no. 27 (2020): 2603–15; L.R. Baden, H.M. El Sahly, B. Essink, et al., "Efficacy and Safety of the mRNA-1273 SARS-CoV-2 Vaccine," *NEJM* 384, no. 5 (2020).

51 Hiam Chemaitelly, Patrick Tang, Mohammad R. Hasan, et al., "Waning of BNT162b2 Vaccine Protection against SARS-CoV-2 Infection in Qatar," *NEJM* 385, no. 24 (2021): e83, published online October 6, 2021, https://www.nejm.org/doi/full/10.1056/NEJMoa2114114.

52 Grace Sparks, Marley Presiado, Isabelle Valdes, Ashley Kirzinger, and Mollyann Brodie, "KFF COVID-19 Vaccine Monitor: March 2023," April 3, 2023, https://www.kff.org/coronavirus-covid-19/poll-finding/kff-covid-19-vaccine-monitor-march-2023/.

53 "People with Certain Medical Conditions and COVID-19 Risk Factors," CDC—COVID-19, June 24, 2024, https://www.cdc.gov/covid/risk-factors/index.html.

54 "Types of COVID-19 Treatment," CDC, July 12, 2024, https://www.cdc.gov/covid/treatment/index.html.

55 "Influenza Antiviral Medications: Summary for Clinicians," CDC, December 8, 2023, https://www.cdc.gov/flu/hcp/antivirals/summary-clinicians.html.

56 Stanley Perlman, MD, PhD, interview with the authors, May 9, 2023.

57 "Global Infectious Disease Therapeutics Market by Type (Drugs,

Vaccines), by Application (HIV/AIDS, Influenza), by Geographic Scope and Forecast," October 2024, Verified Market Reports.

58 "Global Oncology Drug Market by Type (Chemotherapy, Targeted Therapy), by Application (Blood Cancer, Breast Cancer), by Geographic Scope and Forecast," March 2024, Verified Market Reports.

59 "J&J's Janssen to Close Part of Its Vaccine Division–Telegraaf," Reuters, August 24, 2023, https://www.reuters.com/business/healthcare -pharmaceuticals/jjs-janssen-close-part-its-vaccine-division-2023 -08-23/.

60 Hannah Kuchler and Kana Inagaki, "Japanese Drugmaker Urges G7 to Fix Infectious Diseases Market," *Financial Times,* August 28, 2023.

61 "Novo Nordisk Foundation, Open Philanthropy, and Bill & Melinda Gates Foundation Launch Initiative to Support New Antiviral Medicines for Future Pandemics," Gates Foundation press release, March 14, 2022, https://www.gatesfoundation.org/ideas/media-center/press-releases /2022/03/funding-new-antiviral-medicines-and-preventing-future -pandemics.

62 "$20 Million Boost for Development of Antiviral Drugs against Pandemic Influenza," Novo Nordisk Foundation, news release, May 27, 2024, https: //padinitiative.com/announcements/.

63 Max Kozlov, "What Happened to the 'Game Changing' Covid Drug Paxlovid?," *Nature* 613 (2023): 224–25.

64 "COVID Data Tracker: Trends in United States COVID-19 Deaths, Emergency Department (ED) Visits, and Test Positivity by Geographic Area," CDC, https://covid.cdc.gov/covid-data-tracker/#trends _totaldeaths_select_00.

65 Jennifer Hammond, Heidi Leister-Tebbe, Annie Gardner, et al., "Oral Nirmatrelvir for High-Risk, Nonhospitalized Adults with Covid-19," *NEJM* 386 (2022): 1397–408.

66 Juan Jiang, Yantong Li, Qiaoling Jiang, Yu Jiang, Hongqian Qin, and Yuanyuan Li, "Early Use of Oral Antiviral Drugs and the Risk of Post COVID-19 Syndrome: A Systematic Review and Network Meta-analysis," *Journal of Infection* 89, no. 2 (2024): 106190.

67 "FDA Approves First Oral Antiviral for Treatment of COVID-19 in Adults," US Food and Drug Administration, May 25, 2023, https://www .fda.gov/news-events/press-announcements/fda-approves-first-oral -antiviral-treatment-covid-19-adults.

68 Tegan K. Boehmer, Emily H. Koumans, Elizabeth L. Skillen, et al., "Racial and Ethnic Disparities in Outpatient Treatment of COVID19— United States, January–July 2022," *MMWR* 71, no. 43 (2022): 1359–65.

69 "Just a Quarter of Pfizer's COVID-19 Treatment Orders Will Go to Developing Countries," OXFAM International, November 21, 2022, https:

//www.oxfam.org/en/press-releases/just-quarter-pfizers-covid-19
-treatment-orders-will-go-developing-countries.

70 Justin S. Lee, Jason M. Goldstein, Jonathan L. Moon, et al., "Analysis of
the Initial Lot of the CDC 2019-Novel Coronavirus (2019-nCoV) Real-
Time RT-PCR Diagnostic Panel," *PLOS One* 16, no. 12 (2021): e0260487.

71 BinaxNOW COVID-19 Ag Card/Product Insert, Abbott, https://www.fda
.gov/media/141570/download.

72 Katie Thomas and Natasha Singer, "FDA Authorizes First In-Home Test
for Coronavirus," *New York Times*, April 21, 2020, updated May 8, 2020,
https://www.nytimes.com/2020/04/21/health/fda-in-home-test
-coronavirus.html?searchResultPosition=1.

73 "U.S. FDA to Review Fewer Emergency Use Requests for COVID Tests,"
Reuters, September 27, 2022.

74 Riley Griffin and Amelia Pollard, "COVID Test Maker Once Crucial to
US Response Set to Liquidate after Failed Sale," Bloomberg, June 14,
2023; Amelia Pollard, "COVID Test Maker Lucira Goes Bankrupt as
Demand for Kits Wanes," Bloomberg, February 22, 2023.

CHAPTER FIVE

1   Mari Eccles, "Hydroxychloroquine Could Have Caused 17,000 Deaths
during Covid, Study Finds," Politico, January 4, 2024.

2   "Study: Nearly 17,000 Deaths Linked to Hydroxychloroquine during
Early Covid," *KFF Health News,* January 5, 2024, https://kffhealthnews
.org/morning-breakout/study-nearly-17000-deaths-linked-to-
hydroxycholoroquine-during-early-covid/.

3   National Association of County and City Health Officials Directory of
Local Health Departments, https://www.naccho.org/membership/lhd
-directory?searchType=standard&lhd-search=health&lhd-state=#card
-filter.

4   "Risk Communication and Community Engagement," WHO, https:
//www.who.int/emergencies/risk-communications.

5   "Trump Tells Woodward He Deliberately Downplayed Coronavirus
Threat," NPR, *Morning Edition*, September 10, 2020, https://www.npr.org
/2020/09/10/911368698/trump-tells-woodward-he-deliberately
-downplayed-coronavirus-threat.

6   Shawn Boburg, Robert O'Harrow Jr., Neena Satija, and Amy Goldstein,
"Inside the Coronavirus Testing Failure: Alarm and Dismay among the
Scientists Who Sought to Help," *Washington Post*, April 3, 2020.

7   *Meet the Press,* NBC News, transcript, January 31, 2021, accessed January
27, 2023, https://www.nbcnews.com/meet-the-press/meet-press-january
-31-2021-n1256299.

8   "COVID Data Tracker: Trends in United States COVID-19 Deaths,
Emergency Department (ED) Visits, and Test Positivity by Geographic

Area," CDC, https://covid.cdc.gov/covid-data-tracker/#trends
_totaldeaths_select_00.

9   "Coronavirus in the U.S.: Latest Map and Case Count," *New York Times,*
     March 23, 2023, https://www.nytimes.com/interactive/2021/us/covid
     -cases.html.

10   Winston Churchill, "The Bright Gleam of Victory: Speech at the Lord
     Mayor's Day Luncheon at the Mansion House, London, 10 November
     1942," transcript, International Churchill Society, https://winston
     churchill.org/resources/speeches/1941-1945-war-leader/the-bright
     -gleam-of-victory/.

11   A. Lu-Culligan and A. Iwasaki, "The False Rumors about Vaccines That
     Are Scaring Women," *New York Times,* January 26, 2021; Jennifer Abbasi,
     "Widespread Misinformation about Infertility Continues to Create
     COVID-19 Vaccine Hesitancy," *JAMA* 327, no. 11 (2022): 1013–15.

12   Peter M. Sandman, "Commentary: 8 Things US Pandemic
     Communicators Still Get Wrong," CIDRAP, December 9, 2021, https:
     //www.cidrap.umn.edu/commentary-8-things-us-pandemic-communicators
     -still-get-wrong.

13   Sandman, "Commentary: 8 Things."

14   Elizabeth Cooney, "'I'm Deeply Concerned': Francis Collins on Trust in
     Science, How Covid Communications Failed, and His Current
     Obsession," *STAT News,* September 19, 2022.

15   Sandman, "Commentary: 8 Things."

16   Oliver Willis, "Trump Says He Lied about Virus Threat Because He's a
     'Cheerleader' for America," *American Journal News,* September 9, 2020.

17   Sandman, "Commentary: 8 Things."

18   Sandra C. Quinn, PhD, interview with the authors, January 26, 2023.

19   Tracey E. Wilson, Marilyn Fraser-White, Kim M. Williams, et al.,
     "Barbershop Talk with Brothers: Using Community-Based Participatory
     Research to Develop and Pilot Test a Program to Reduce HIV Risk among
     Black Heterosexual Men," *AIDS Education and Prevention* 26, no. 5 (2014):
     383–97.

20   The Great Barrington Declaration (website), https://gbdeclaration.org
     /#read.

21   Apoorva Mandavilli and Sheryl Gay Stolberg, "A Viral Theory Cited by
     Health Officials Draws Fire from Scientists," *New York Times,* October 19,
     2020, updated October 23, 2020, https://www.nytimes.com/2020/10/19
     /health/coronavirus-great-barrington.html.

22   Tedros Adhanom Ghebreyesus, "WHO Director-General's Opening
     Remarks at the Media Briefing on COVID-19 — 12 October 2020,"
     transcript, Speeches: Detail, WHO, https://www.who.int/director-general
     /speeches/detail/who-director-general-s-opening-remarks-at-the-media
     -briefing-on-covid-19---12-october-2020.

23   Iván Martínez-Baz, Ana Miqueleiz, Nerea Egüés et al., "Effect of COVID-19 Vaccination on the SARS-CoV-2 Transmission among Social and Household Close Contacts: A Cohort Study," *Journal of Infection and Public Health* 16 (2023): 410–17.

24   Sandman, "Commentary: 8 Things."

25   Dorey Scheimer and Meghna Chakrabarti, "Why the U.S. Needs a Reckoning on Lockdowns before the Next Pandemic," WBUR, *On Point*, June 11, 2024, https://www.wbur.org/onpoint/2024/06/11/pandemic -covid-lockdown-public-health-physician; Wesley J. Smith, "Francis Collins Disappointed as a Public-Health Leader," *National Review*, December 30, 2023.

26   Peter M. Sandman, "The CDC Tells More of the Truth than Usual about Flu Vaccine Effectiveness," Peter Sandman Risk Communication (website), 2013, https://www.psandman.com/col/presser.htm.

27   Michael T. Osterholm, Nicholas S. Kelley, Alfred Sommer, and Edward A. Belongia, "Efficacy and Effectiveness of Influenza Vaccines: A Systematic Review and Meta-analysis," *Lancet Infectious Diseases* 12, no. 1 (2012), epub October 25, 2011. Erratum in *Lancet Infectious Diseases* 12, no. 9 (2012): 655.

28   Hannah Chung, Michael A. Campitelli, Sarah A. Buchan, et al., "Measuring Waning Protection from Seasonal Influenza Vaccination during Nine Influenza Seasons, Ontario, Canada, 2010/11 to 2018/19," *Eurosurveillance* 29, no. 8 (2024): pii=2300239.

29   Sami Al Hajjar and Kenneth McIntosh, "The First Influenza Pandemic of the 21st Century," *Annals of Saudi Medicine* 30, no. 1 (2010): 1–10.

30   Kristine A. Moore, Marc Lipsitch, John M. Barry, and Michael T. Osterholm, "COVID-19: The CIDRAP Viewpoint—Part 1: The Future of the COVID-19 Pandemic: Lessons Learned from Pandemic Influenza," April 30, 2020, https://www.cidrap.umn.edu/sites/default/files /downloads/cidrap-covid19-viewpoint-part1_0.pdf.

31   Moore et al., "COVID-19: The CIDRAP Viewpoint—Part 1."

32   Eric J. Topol, "Coronavirus Variants Don't Have to Be Scary. Still, Mask Up," *New York Times*, April 13, 2021.

33   Michael T. Osterholm and Mark Olshaker, "Facing Covid-19 Reality: A National Lockdown Is No Cure," *Washington Post*, March 21, 2020.

34   Sandman, "Commentary: 8 Things."

35   Fernando P. Polack, Stephen J. Thomas, Nicholas Kitchin, et al., "Safety and Efficacy of the BNT162b2 mRNA Covid-19 Vaccine," *NEJM* 383, no. 27 (2020): 2603–15; L.R. Baden, H.M. El Sahly, B. Essink, et al., "Efficacy and Safety of the mRNA-1273 SARS-CoV-2 Vaccine," *NEJM* 384, no. 5 (2020).

36   Sandman, "Commentary: 8 Things."

37 Sandman, "Commentary: 8 Things."

38 Sandman, "Commentary: 8 Things."

39 Peter J. Hotez and Arthur L. Caplan, *Vaccines Did Not Cause Rachel's Autism: My Journey as a Vaccine Scientist, Pediatrician, and Autism Dad* (Baltimore: Johns Hopkins University Press, 2018).

40 Joe Rogan, host, *The Joe Rogan Experience,* podcast, episode 1439, "Michael Osterholm," March 10, 2020.

41 Joe Rogan, host, *The Joe Rogan Experience,* podcast, episode 1779, "Michael Osterholm," February 18, 2022.

42 Nate Silver (@NateSilver538), "Check out, for example, this interview with Dr. Michael Osterholm on Meet The Press last week. It's relentlessly negative," Twitter, April 12, 2021, 7:34 a.m.

43 *Meet the Press,* April 4, 2021, NBC News, transcript, 13–15, https://www.nbcnews.com/meet-the-press/meet-press-april-04-2021-n1262993.

44 Lauren Leatherby, "As Variants Have Spread, Progress against the Virus in U.S. Has Stalled," *New York Times,* April 6, 2021.

45 Nate Rattner and Rich Mendez, "Covid Cases Are Rising Again in All 50 States across U.S. as Delta Variant Tightens Its Grip," CNBC, July 23, 2021, https://www.cnbc.com/2021/07/23/covid-cases-are-rising-again-in-all-50-states-across-us-as-delta-variant-tightens-its-grip.html.

46 Lauren Leatherby, Charlie Smart, and Amy Schoenfeld Walker, "Omicron Drives U.S. Virus Cases Past Delta's Peak, *New York Times,* December 23, 2021.

47 Aamer Madhani, "Herman Cain Treated for COVID-19 after Attending Trump Rally," PBS, July 2, 2020, https://www.pbs.org/newshour/health/herman-cain-treated-for-covid-19-after-attending-trump-rally; Aimee Ortiz and Katharine Q. Seelye, "Herman Cain, Former C.E.O. and Presidential Candidate, Dies at 74," *New York Times,* July 30, 2020, updated August 3, 2020, https://www.nytimes.com/2020/07/30/us/politics/herman-cain-dead.html.

48 Meredith Deliso and Joshua Hoyos, "Minnesota Sees No Rise in COVID-19 Cases Tied to Protests: Health Official," ABC News, June 22, 2020, https://abcnews.go.com/US/minnesota-sees-rise-covid-19-cases-tied-protests/story?id=71393938.

49 Clara Suñer, Ermengol Coma, Dan Ouchi, Eduardo Hermosilla, Bàrbara Baro, Miquel Àngel Rodríguez-Arias, Jordi Puig, Bonaventura Clotet, Manuel Medina, and Oriol Mitjà, "Association between Two Mass-Gathering Outdoor Events and Incidence of SARS-CoV-2 Infections during the Fifth Wave of COVID-19 in North-East Spain: A Population-Based Control-Matched Analysis," *Lancet Regional Health—Europe* 15 (2022): 100337.

50 "About the U.S. Public Health Service Untreated Syphilis Study at Tuskegee," CDC, September 4, 2024, https://www.cdc.gov/tuskegee /about/index.html.

51 Quinn interview.

52 Nathan Ballantyne, "Epistemic Trespassing," *Mind* 128, no. 510 (2019): 367–95.

53 "Mehdi Hasan Questions Dr. Monica Gandhi on Her Covid-19 Predictions," *Mehdi Hasan Show,* MSNBC, February 4, 2022, https://www .youtube.com/watch?v=VaSTb5kNT4s.

54 Meghan Tocci, "Smartphone History and Evolution," *Beyond Texting* (blog), June 27, 2024, accessed November 12, 2024, https://simpletexting .com/blog/where-have-we-come-since-the-first-smartphone/.

55 Jonathan Mahler, "The White and Gold (No, Blue and Black!) Dress That Melted the Internet," *New York Times,* February 27, 2015.

56 David J. Rothkopf, "When the Buzz Bites Back," *Washington Post,* May 11, 2003.

57 *Novel Coronavirus (2019-nCoV) Situation Report — 13,* WHO, February 2, 2020, https://www.who.int/docs/default-source/coronaviruse/situation -reports/20200202-sitrep-13-ncov-v3.pdf.

58 "Countering Misinformation about COVID-19," WHO, May 11, 2020, updated May 13, 2020, https://www.who.int/news-room/feature-stories /detail/countering-misinformation-about-covid-19.

59 Derek Thompson, "The Pandemic's Wrongest Man," *The Atlantic,* April 1, 2021.

60 Stephanie Alice Baker, Eugene McLaughlin, and Chris Rojek, "Simple Solutions to Wicked Problems: Cultivating True Believers of Anti-vaccine Conspiracies during the COVID-19 Pandemic," *European Journal of Cultural Studies* 27, no. 4 (2023), https://doi.org/10.1177 /13675494231173536.

61 "How to Report Misinformation Online," WHO, 2024, https://www.who .int/campaigns/connecting-the-world-to-combat-coronavirus/how-to -report-misinformation-online.

62 Jennifer Aaker and Victoria Chang, "Obama and the Power of Social Media and Technology," Stanford Graduate School of Business, case no. M321, 2009, https://www.gsb.stanford.edu/faculty-research/case-studies /obama-power-social-media-technology.

63 Nils C. Köbis, Barbora Doležalová, and Ivan Soraperra, "Fooled Twice: People Cannot Detect Deepfakes but Think They Can," *iScience* 24, no. 11 (2021): 103364, https://phys.org/news/2023-09-humans-easily-deepfakes .html#google_vignette.

64 Jamie A. Teixeira da Silva, Panagiotis Tsigaris, and Mohammadamin Erfanmanesh, "Publishing Volumes in Major Databases Related to Covid-19," *Scientometrics* 126 (2021): 831–42.

CHAPTER SIX

1 "Introduction to Public Health Surveillance," Public Health 101 Series presentation, Center for Surveillance, Epidemiology, and Laboratory Services, Division of Scientific Education and Professional Development, CDC, slide 25, https://www.cdc.gov/training-publichealth101/media/pdfs/introduction-to-surveillance.pdf.

2 Eleni Tsiompanou and Spyros G. Marketos, "Hippocrates: Timeless Still," *Journal of the Royal Society of Medicine* 106, no. 7 (2013): 288–92.

3 Stephen B. Thacker and Donna F. Stroup, "Origins and Progress in Surveillance Systems," in *Infectious Disease Surveillance,* 2nd ed., ed. Nkuchia M. M'ikanatha, Ruth Lynfield, Chris A. Van Beneden, and Henriette de Valk, chaps. 2, 21 (Hoboken, NJ: John Wiley & Sons, 2013).

4 Thacker and Stroup, "Origins and Progress."

5 Thacker and Stroup, "Origins and Progress," 22.

6 Henry Connor, "John Graunt F.R.S. (1620–74): The Founding Father of Human Demography, Epidemiology and Vital Statistics," *Journal of Medical Biology* 32, no. 1 (2024): 57–69.

7 Wilson G. Smillie, "Morbidity Reporting...the Basis of Communicable Disease Control," *Public Health Reports* 67, no. 3 (1952): 287.

8 Institute of Medicine (US) Committee for the Study of the Future of Public Health, "A History of the Public Health System," *The Future of Public Health* (Washington, DC: National Academies Press, 1988), 4.

9 Thacker and Stroup, "Origins and Progress," 25.

10 John M. Eyler, "William Farr on the Cholera: The Sanitarian's Disease Theory and the Statistician's Method," *Journal of the History of Medicine* 28, no. 2 (1973).

11 Thacker and Stroup, "Origins and Progress."

12 Thacker and Stroup, "Origins and Progress," 22.

13 Thacker and Stroup, "Origins and Progress," 25.

14 Thacker and Stroup, "Origins and Progress," 26.

15 Thacker and Stroup, "Origins and Progress," 26–27.

16 "Heroes of Public Health: Alexander Langmuir, MD, MPH," Johns Hopkins Bloomberg School of Public Health, https://publichealth.jhu.edu/about/history/heroes-of-public-health/alexander-langmuir-md-mph.

17 William H. Foege, "Alexander D. Langmuir—His Impact on Public Health," *American Journal of Epidemiology* 144, no. 8, supp. (1996): S11–S15.

18 National Association of County and City Health Officials Directory of Local Health Departments, https://www.naccho.org/membership/lhd-directory?searchType=standard&lhd-search=health&lhd-state=#card-filter.

19 Owen Dyer, "Florida Loses Legal Battle to Keep Covid Data Secret," *BMJ* 383 (2023): 2419.

20 Members of the Council of State and Territorial Epidemiologists, interview with the authors, January 3, 2024.

21 Betsy Ladyzhets, "The U.S. Still Doesn't Know How to Track a Pandemic," FiveThirtyEight, March 23, 2022, https://fivethirtyeight.com/features /the-u-s-still-doesn't-know-how-to-track-a-pandemic/.

22 Ladyzhets, "The U.S. Still Doesn't Know How to Track a Pandemic."

23 Whet Moser and Conor Kelly, "To Understand the US Pandemic, We Need Hospitalization Data—and We Almost Have It. The COVID Tracking Project," *The Atlantic*, July 4, 2020, https://covidtracking.com /analysis-updates/hospitalization-data.

24 "Contact Tracing," David J. Sencer CDC Museum Public Health Academy, https://www.cdc.gov/museum/pdf/cdcm-pha-stem-lesson-contact-tracing -lesson.pdf.

25 "Measles (Rubeola)—Measles Symptoms and Complications," CDC, May 9, 2024, https://www.cdc.gov/measles/signs-symptoms/index.html.

26 "Measles (Rubeola)—How Measles Spreads," CDC, April 18, 2024, https: //www.cdc.gov/measles/causes/index.html.

27 "Smallpox—Q&A," WHO, June 28, 2016, https://www.who.int/news-room /questions-and-answers/item/smallpox.

28 F. Fenner, D.A. Henderson, I. Arita, Z. Ježek, and I.D. Ladnyi, *Smallpox and Its Eradication* (Geneva: WHO, 1988), 65, https://iris.who.int /handle/10665/39485.

29 Stephen A. Lauer, Kyra H. Grantz, Qifang Bi, Forrest K. Jones, Qulu Zheng, Hannah R. Meredith, Andrew S. Azman, Nicholas G. Reich, and Justin Lessler, "The Incubation Period of Coronavirus Disease 2019 (COVID-19) from Publicly Reported Confirmed Cases: Estimation and Application," *Annals of Internal Medicine* 172, no. 9 (2020): 577–82.

30 Yu Wu, Liangyu Kang, Zirui Guo, Jue Liu, Min Liu, and Wannian Liang, "Incubation Period of COVID-19 Caused by Unique SARS-CoV-2 Strains: A Systematic Review and Meta-analysis," *JAMA Network Open* 5, no. 8 (2022): e2228008.

31 Sandy Y. Joung, Joseph E. Ebinger, Nancy Sun, et al., "Awareness of SARS-CoV-2 Omicron Variant Infection among Adults with Recent COVID-19 Seropositivity," *JAMA Network Open* 5, no. 8 (2022): e2227241.

32 "Covid-19 Contact Tracing Playbook," Resolve to Save Lives (an initiative of Vital Strategies), December 21, 2020, https://contacttracingplaybook .resolvetosavelives.org/.

33 Kristine Moore, Jill DeBoer, Richard Hoffman, Patrick McConnon, Dale Morse, and Michael Osterholm, "COVID-19: The CIDRAP Viewpoint—Part 4: Contact Tracing for COVID-19: Assessing Needs, Using a Tailored Approach," reviewed by Jeffery Duchin, June 2, 2020, https://www.cidrap .umn.edu/sites/default/files/downloads/cidrap-covid19-viewpoint-part4.pdf.

34   Selena Simmons-Duffin, "States Nearly Doubled Plans for Contact
     Tracers since NPR Surveyed Them 10 Days Ago," NPR, *Morning Edition*,
     May 7, 2020, https://www.npr.org/sections/health-shots/2020/04/28
     /846736937/we-asked-all-50-states-about-their-contact-tracing-capacity
     -heres-what-we-learne.

35   *Health Departments: Interim Guidance on Developing a COVID-19 Case
     Investigation & Contact Tracing Plan,* US Department of Health and
     Human Services, CDC, CS317074-A, June 1, 2020, https://stacks.cdc.gov
     /view/cdc/88623/cdc_88623_DS1.pdf.

36   R. Ryan Lash, Patrick K. Moonan, Brittany L. Byers, et al., "COVID-19
     Case Investigation and Contact Tracing in the US, 2020," *JAMA Network
     Open* 4, no. 6 (2021): e2115850; Seonghye Jeon, Gabriel Rainisch, R. Ryan
     Lash, Patrick K. Moonan, John E. Oeltmann, Bradford Greening Jr.,
     Bishwa B. Adhikari, Contact Tracing Impact Group, and Martin I.
     Meltzer, "Estimates of Cases and Hospitalizations Averted by COVID-19
     Case Investigation and Contact Tracing in 14 Health Jurisdictions in the
     United States," *Journal of Public Health Management and Practice* 28, no. 1
     (2022): 16–24.

37   "Training and Workforce Development for COVID-19 Case Investigation
     and Contact Tracing," CDC, CS 328076-B, May 31, 2022, https://archive
     .cdc.gov/www_cdc_gov/coronavirus/2019-ncov/php/contact-tracing/pdf
     /TrainingContactTracingWorkforce_508.pdf.

38   Denis Mongin, Nils Bürgisser, the Covid-SMC Study Group, and Delphine
     Sophie Courvoisier, "Time Trends and Modifiable Factors of COVID-19
     Contact Tracing Coverage, Geneva, Switzerland, June 2020 to February
     2022," *Eurosurveillance* 29, no. 3 (2024): 2300228.

39   "State Approaches to Contact Tracing during the COVID-19 Pandemic,"
     National Academy for State Health Policy, September 23, 2022, https:
     //nashp.org/state-tracker/state-approaches-to-contact-tracing-during
     -the-covid-19-pandemic/.

40   "Public Health Agencies Transitioning Away from Universal Case
     Investigation and Contact Tracing for Individual Cases of COVID-19,"
     Joint Statement of the Association of Public Health Laboratories,
     Association of State and Territorial Health Officials, Big Cities Health
     Coalition, Council of State and Territorial Epidemiologists, and National
     Association of County and City Health Officials, January 24, 2022, https:
     //www.naccho.org/uploads/downloadable-resources/CICT_Partner
     _Statement_01_24_2022.pdf.

41   Natalie Dean, "Tracking COVID-19 Infections: Time for Change," *Nature*,
     February 8, 2022.

42   "Real-time Assessment of Community Transmission (REACT) Study—
     About the REACT Programme," Imperial College London, https:

//www.imperial.ac.uk/medicine/research-and-impact/groups/react
-study/.

43   Paul Elliott, Matthew Whitaker, David Tang, et al., "Design and
     Implementation of a National SARS-CoV-2 Monitoring Program in
     England: REACT-1 Study," *American Journal of Public Health* 113, no. 5
     (2023): 545–54.

44   Marc Lipsitch, DPhil, interview with the authors, January 16, 2024.

45   James D. Trask and John R. Paul, with the technical assistance of
     John T. Riordan, "Periodic Examination of Sewage for the Virus of
     Poliomyelitis," *Journal of Experimental Medicine* 75, no. 1 (1942): 1–6.

46   Life in the Lab Staff, "A Brief History of Wastewater Testing and
     Pathogen Detection," *ThermoFisher Scientific* (blog), October 5, 2021,
     https://www.thermofisher.com/blog/life-in-the-lab/a-brief-history-of
     -wastewater-testing-and-pathogen-detection/.

47   Gertjan Medema, Leo Heijnen, Goffe Elsinga, Ronald Italiaander, and
     Anke Brouwer, "Presence of SARS-Coronavirus-2 RNA in Sewage and
     Correlation with Reported COVID-19 Prevalence in the Early Stage of the
     Epidemic in the Netherlands," *Environmental Science and Technology Letters*
     7, no. 7 (2020): 511–16.

48   "About CDC's National Wastewater Surveillance System (NWSS)," CDC,
     November 6, 2024, https://www.cdc.gov/nwss/about.html#print.

49   Carly Adams, Megan Bias, Rory M. Welsh, Jenna Webb, Heather Reese,
     Stephen Delgado, John Person, Rachel West, Soo Shin, and Amy Kirby,
     "The National Wastewater Surveillance System (NWSS): From Inception
     to Widespread Coverage, 2020–2022, United States," *Science of the Total
     Environment* 924 (2024): 171566, http://doi.org/10.1016/j.scitotenv.2024
     .171566.

50   "Traveler-Based Genomic Surveillance for Early Detection of New SARS-
     CoV-2 Variants," CDC, June 17, 2024, https://wwwnc.cdc.gov/travel/page
     /travel-genomic-surveillance.

51   "Promoting Interoperability Programs," Centers for Medicare &
     Medicaid Services (CMS), November 5, 2024, https://www.cms.gov
     /medicare/regulations-guidance/promoting-interoperability-programs.

52   "Interim Final Rule (IFC), CMS-3401-IFC; Requirements and
     Enforcement Process for Reporting of COVID-19 Data Elements for
     Hospitals and Critical Access Hospitals," QSO-21-03-Hospitals/CAHs,
     CMS, October 6, 2020, https://www.cms.gov/medicareprovider
     -enrollment-and-certificationsurveycertificationgeninfopolicy-and
     -memos-states-and/interim-final-rule-ifc-cms-3401-ifc-requirements
     -and-enforcement-process-reporting-covid-19-data.

53   Andy Slavitt, MBA, interview with the authors, January 24, 2024.

54   Slavitt interview.

55 Slavitt interview.

56 Marc Lipsitch, Mary T. Bassett, John S. Brownstein et al., "Infectious Disease Surveillance Needs for the United States: Lessons from Covid-19," *Frontiers in Public Health* 12 (2024): 1408193.

57 Slavitt interview.

58 Kristy Snyder, "Cerner vs. Epic EHR (2024 Comparison)," Forbes Advisor, April 29, 2024, https://www.forbes.com/advisor/business/software/cerner -vs-epic/.

59 Slavitt interview.

60 Lipsitch interview.

61 "Data Modernization Initiative," CDC, October 18, 2024, accessed November 17, 2024, https://www.cdc.gov/data-modernization/php /about/dmi.html.

CHAPTER SEVEN

1 Dave Roos, "Woodrow Wilson Got the Flu in a Pandemic during the World War I Peace Talks," *History,* October 6, 2020, https://www.history .com/news/woodrow-wilson-1918-pandemic-world-war-i.

2 Quint Forgey and Matthew Choi, "'This Is Deadly Stuff': Tapes Show Trump Acknowledging Virus Threat in February," Politico, September 9, 2020.

3 "Trump Tells Woodward He Deliberately Downplayed Coronavirus Threat," NPR, *Morning Edition,* September 10, 2020, https://www.npr .org/2020/09/10/911368698/trump-tells-woodward-he-deliberately -downplayed-coronavirus-threat.

4 "Trump Tells Woodward."

5 Winston Churchill, "Blood, Toil, Tears and Sweat: First Speech as Prime Minister to House of Commons," May 13, 1940, transcript, International Churchill Society, https://winstonchurchill.org/resources/speeches /1940-the-finest-hour/blood-toil-tears-sweat/.

6 Edward R. Murrow, *In Search of Light: The Broadcasts of Edward R. Murrow 1938–1961,* ed. Edward Bliss Jr. (New York: Alfred A. Knopf, 1967).

7 "Proclamation on Declaring a National Emergency Concerning the Novel Coronavirus Disease (COVID-19) Outbreak," March 13, 2020, https: //trumpwhitehouse.archives.gov/presidential-actions/proclamation -declaring-national-emergency-concerning-novel-coronavirus-disease -covid-19-outbreak/.

8 "CARES Act," Office of Inspector General, January 15, 2021, https://oig .treasury.gov/cares-act.

9 Keith Zubrow, "Operation Warp Speed Preps for COVID-19 Vaccine Delivery upon FDA Approval," *60 Minutes Overtime,* CBS News, November 8, 2020, https://www.cbsnews.com/news/operation-warp -speed-covid-19-vaccine-delivery-fda-60-minutes-2020-11-08/.

10   "Explaining Operation Warp Speed," US Department of Health and Human Services, https://cnrk.cnic.navy.mil/Portals/85/CNFK /Documents/fact-sheet-operation-warp-speed.pdf?ver=zDr-vPH0 _IIqXFfQnWr3gg%3D%3D.

11   Vanessa Romo, "Dr. Scott Atlas, Special Coronavirus Adviser to Trump, Resigns," NPR, November 30, 2020, https://www.npr.org/2020/11/30 /940376041/dr-scott-atlas-special-coronavirus-adviser-to-trump -resigns.

12   Roy H. Perlis, Kristin Lunz Trujillo, Jon Green, Alauna Safarpour, James N. Druckman, Mauricio Santillana, Katherine Ognyanova, and David Lazer, "Misinformation, Trust, and Use of Ivermectin and Hydroxychloroquine for COVID-19," *JAMA Health Forum* 4, no. 9 (2023): e233257.

13   "Coronavirus: Trump's Disinfectant and Sunlight Claims Fact-Checked," Reality Check team, BBC News, April 24, 2020, https://www.bbc.com /news/world-us-canada-52399464.

14   "Trump Administration Interfered with CDC's Public Outreach on COVID-19," Union of Concerned Scientists, March 17, 2022.

15   Philip Zelikow and the Commission Planning Group, *Lessons from the Covid War: An Investigative Report* (New York: PublicAffairs, 2023), 130–31.

16   "President Biden Announces American Rescue Plan," the White House, January 20, 2021, https://www.whitehouse.gov/briefing-room/legislation /2021/01/20/president-biden-announces-american-rescue-plan/.

17   David Cohen and Adam Cancryn, "Biden on '60 Minutes': 'The Pandemic Is Over,'" Politico, September 18, 2022.

18   "COVID Data Tracker: Trends in United States COVID-19 Deaths, Emergency Department (ED) Visits, and Test Positivity by Geographic Area," CDC, https://covid.cdc.gov/covid-data-tracker/#trends _totaldeaths_select_00.

19   Betsy Ladyzhets, "NIH Documents Show How $1.6 Billion Long Covid Initiative Has Failed So Far to Meet Its Goals," *STAT News,* May 31, 2024.

20   "Dr. Peter Hotez on the Anti-science Movement and Declining Joe Rogan's Debate Challenge," AMA Update, July 13, 2023, https://www .ama-assn.org/delivering-care/public-health/dr-peter-hotez-anti-science -movement-and-declining-joe-rogan-s-debate.

21   "Advocacy: New Laws & Regulations," TNP [Texas Nurse Practitioners], https://texasnp.org/news-laws-and-regulations/; Ciara McCarthy, "Reality Check: Doctors Outraged after Texas Bans Health Departments from Promoting COVID Vaccines," *Fort Worth Star-Telegram,* March 21, 2024, https://www.star-telegram.com/news/politics-government/state -politics/article286604225.html.

22   "Updated Guidance for COVID-19 Boosters for the Fall and Winter 2024– 2025 Season," Florida Department of Health, September 12, 2024,

September 13, 2024, https://www.floridahealth.gov/newsroom/2024/09/20210912-UpdatedGuidanceCOVID-19.html.

23 Katherine M. Jia, William P. Hanage, Marc Lipsitch, Amelia G. Johnson, Avnika B. Amin, Akilah R. Ali, Heather M. Scobie, and David L. Swerdlow, "Estimated Preventable COVID-19-Associated Deaths Due to Non-vaccination in the United States," *European Journal of Epidemiology* 38 (2023): 1125–28.

24 María L. Avila-Aguero, Kattia Camacho-Badilla, and Rolando Ulloa-Gutierrez, "Measles Outbreaks: What Does It Represent for the Elimination Strategy in the Region of the Americas? A Call for the Action," *Expert Review of Vaccines* 14, no. 8 (2015): 1043–45.

25 Selena Simmons-Duffin and Koko Nakajima, "This Is How Many Lives Could Have Been Saved with COVID Vaccinations in Each State," NPR, May 13, 2022, https://www.npr.org/sections/health-shots/2022/05/13/1098071284/this-is-how-many-lives-could-have-been-saved-with-covid-vaccinations-in-each-sta.

26 Peter J. Hotez, *The Deadly Rise of Anti-Science: A Scientist's Warning* (Baltimore: Johns Hopkins University Press, 2023), 20.

27 C. Dominik Güss, Lauren Boyd, Kelly Perniciaro, Danielle C. Free, J.R. Free, and Ma. Teresa Tuason, "The Politics of COVID-19: Differences between U.S. Red and Blue States in COVID-19 Regulations and Deaths," *Health Policy OPEN* 5 (2023): 100107.

28 Ashley Kirzinger, Grace Sparks, Isabelle Valdes, and Liz Hamel, "KFF COVID-19 Vaccine Monitor September 2023: Partisanship Remains Key Predictor of Views of COVID-19, Including Plans to Get Latest COVID-19 Vaccine," KFF, September 27, 2023, https://www.kff.org/health-misinformation-and-trust/poll-finding/kff-covid-19-vaccine-monitor-september-2023/.

29 Jacob Wallace, Paul Goldsmith-Pinkham, and Jason L. Schwartz, "Excess Death Rates for Republican and Democratic Registered Voters in Florida and Ohio during the COVID-19 Pandemic," *JAMA Internal Medicine* 183, no. 9 (2023): 916–23.

30 David A. Asch, Chongliang Luo, and Yong Chen, "Reports of COVID-19 Vaccine Adverse Events in Predominantly Republican vs Democratic States," *JAMA Network Open* 7, no. 3 (2024): e244177.

31 Asch et al., "Reports of COVID-19 Vaccine Adverse Events."

32 Sandro Galea, "Within Reason? Ensuring Public Health Matters in Coming Decades," Gaylord Anderson Memorial Lecture, University of Minnesota, February 27, 2024.

33 Galea, "Within Reason?," slide 39, citing C.R. Hatton et al., "American Trust in Science & Institutions in the Time of COVID-19," *Daedalus: Journal of the American Academy of Arts and Science* 151, no. 4 (2022): 83–97, https://doi.org/10.1162/DAED_a_01945.

34   Galea, "Within Reason?," slide 40, citing Morning Consult National Tracking Poll #2212099, conducted December 14–19, 2022.

35   Galea, "Within Reason?," slide 44.

36   Reese Gorman, "Republican Study Committee Lays Out Public Health Priorities," *Washington Examiner*, August 8, 2023.

37   "Mandate for the Independent Commission (Coronavirus Commission) to Review Lessons Learned from the COVID-19 Pandemic in Norway," Koronakommisjonen, April 2020, https://www.koronakommisjonen.no/mandate-in-english/.

38   MPR News Staff, "'Buckle It Up': Walz Orders MN to 'Stay at Home' to Curb Virus Spread," MPR News, March 25, 2020, https://www.mprnews.org/story/2020/03/25/latest-on-covid19-in-minnesota.

39   Martha Coventry, "Modeling COVID-19 for Minnesota," University of Minnesota School of Public Health News, April 10, 2020, https://www.sph.umn.edu/news/modeling-covid-19-for-minnesota/.

40   Coventry, "Modeling COVID-19 for Minnesota."

41   "Mortality (Death) Data," Minnesota Department of Health, November 21, 2024, https://www.health.state.mn.us/diseases/coronavirus/stats/death.html.

42   Emmanuela Gakidou, "Why Our COVID-19 Total Death Projections for the United States More Than Doubled," *Insights* (blog), IHME, May 20, 2020, https://www.healthdata.org/news-events/insights-blog/acting-data/why-our-covid-19-total-death-projections-united-states-more.

43   Bill Chappell, "Fauci Says U.S. Coronavirus Deaths May Be 'More Like 60,000'; Antibody Tests on Way," NPR, April 9, 2020, https://www.npr.org/2020/04/09/830664814/fauci-says-u-s-coronavirus-deaths-may-be-more-like-60-000-antibody-tests-on-way.

44   Gakidou, "Our COVID-19 Total Death Projections."

45   "New IHME Forecasts Show More Than 200,000 US Deaths by November 1," *PR Newswire*, July 7, 2020.

46   "COVID Data Tracker."

47   "New IHME Forecasts."

48   Robert C. Reiner Jr., Ryan M. Barber, James K. Collins, et al. "Modeling COVID-19 Scenarios for the United States," *Nature Medicine* 27 (2021): 94–105.

49   Michael T. Nietzel, "CDC Awarding More Than $250 Million for Network of 13 Infectious Disease Forecasting Centers," *Forbes*, September 22, 2023.

50   Galea, "Within Reason?," slide 24, citing data from the Bureau of Labor Statistics.

51   Rachael M. Billock, Andrea L. Steege, and Arialdi Miniño, "COVID-19 Mortality by Usual Occupation and Industry: 46 States and New York City, United States, 2020," *National Vital Statistics Reports* 71, no. 6 (2022), https://www.cdc.gov/nchs/data/nvsr/nvsr71/nvsr71-06.pdf.

52   Neel Kashkari, MBA, interview with the authors, October 18, 2022.

53   Jason Furman, PhD, interview with the authors, February 27, 2024.

54   Furman interview.

55   Samuel Hanson, Adi Sunderam, and Eric Zwick, "Business Continuity Insurance in the Next Disaster," Aspen Institute Economic Strategy Group, December 1, 2021, https://www.economicstrategygroup.org /publication/hanson-sunderam-zwick/.

56   Furman interview.

57   Kashkari interview.

58   Furman interview.

59   Furman interview.

60   "Debt Service Suspension Initiative," World Bank Group, March 10, 2020, https://www.worldbank.org/en/topic/debt/brief/covid-19-debt-service -suspension-initiative.

61   "NHE Fact Sheet," CMS, last modified September 10, 2024, https://www .cms.gov/data-research/statistics-trends-and-reports/national-health -expenditure-data/nhe-fact-sheet.

62   Emma Wager, Matthew McGough, Shameek Rakshit, Krutika Amin, and Cynthia Cox, "How Does Health Spending in the U.S. Compare to Other Countries?," Peterson-KFF Health System Tracker, January 23, 2024, https://www.healthsystemtracker.org/chart-collection/health-spending -u-s-compare-countries/.

63   Wager et al., "Health Spending in the U.S."

64   "NHE Fact Sheet."

65   Munira Z. Gunja, Evan D. Gumas, and Reginald D. Williams II, "U.S. Health Care from a Global Perspective, 2022: Accelerating Spending, Worsening Outcomes," Commonwealth Fund Issue Brief, January 31, 2023, https://www.commonwealthfund.org/publications/issue-briefs /2023/jan/us-health-care-global-perspective-2022.

66   Anna M. Cushing, Emily M. Bucholz, Alyna T. Chien, Daniel A. Rauch, and Kenneth A. Michelson, "Availability of Pediatric Inpatient Services in the United States," *Pediatrics* 148, no. 1 (2021): e2020041723.

67   "Lost on the Frontline," *The Guardian*, April 8, 2021, https://www .theguardian.com/us-news/ng-interactive/2020/aug/11/lost-on -the-frontline-covid-19-coronavirus-us-healthcare-workers-deaths -database.

68   Brad Spellberg, MD, interview with the authors, April 26, 2024.

69   Maureen Anthony, "Hospital-at-Home," *Home Healthcare Now,* May/June 2021.

70   Johns Hopkins Medicine, "Hospital at Home® Home-Based Care for Older Adults," https://www.johnshopkinssolutions.com/wp-content /uploads/2018/10/Johns_Hopkins_Hospital-at-Home_Overview.pdf.

71   Eric Topol, MD, interview with the authors, May 10, 2023.

72 Richard Eisenberg, "What to Know about Medicare and Hospital at Home Programs," *Fortune Well,* June 24, 2024, https://fortune.com/well/article/medicare-hospital-at-home-programs/.

73 Spellberg interview.

74 Antony J. Blinken, US Secretary of State, "The Administration's Approach to the People's Republic of China," speech at the George Washington University, Washington, DC, May 26, 2022, https://2021-2025.state.gov/the-administrations-approach-to-the-peoples-republic-of-china/.

75 Spellberg interview.

CHAPTER EIGHT

1 *The National Blueprint for Biodefense: Immediate Action Needed to Defend against Biological Threats,* Report of the Bipartisan Commission on Biodefense, Washington, DC, April 2024, 2, https://biodefensecommission.org/wp-content/uploads/2024/05/National-Blueprint-for-Biodefense-2024_final_digital.pdf.

2 James D. Cherry, "The Chronology of the 2002–2003 SARS Mini Pandemic," *Paediatric Respiratory Reviews* 5, no. 4 (2004): 262–69.

3 Michael T. Osterholm, "Preparing for the Next Pandemic," *Foreign Affairs,* June 1, 2005.

4 "How Did COVID-19 Get It's [*sic*] Name?," CDC, September 1, 2020, https://stacks.cdc.gov/view/cdc/93060.

5 Olivia Webb Kosloff, "A National Defense Strategy for Generic Drugs," *American Affairs* 8, no. 2 (2024): 35–44.

6 David Robson, "Why the Flu of 1918 Was So Deadly," BBC, October 30, 2018, https://www.bbc.com/future/article/20181029-why-the-flu-of-1918-was-so-deadly.

7 Cherry, "Chronology of the 2002–2003 SARS Mini Pandemic."

8 Vanessa Kerry, "The Pandemic Treaty Will Succeed Only with Investment to Build Strong Health Systems," *STAT News,* May 31, 2024, https://www.statnews.com/2024/05/31/pandemic-treaty-investment-build-strong-health-systems/.

9 Lawrence H. Summers, Robert Hecht, and Shan Soe-Lin, "How the Global Pandemic Fund Can Live Up To Its Great Potential," *Washington Post,* January 30, 2023.

10 Celia W. Dugger, "Study Cites Toll of AIDS Policy in South Africa," *New York Times,* November 25, 2008.

11 Carmen Paun, "RFK Jr: 'We're Not Gonna Take Vaccines Away from Anybody,'" Politico, November 6, 2024, https://www.politico.eu/article/robert-f-kennedy-jr-not-taking-vaccines-away-from-anybody/.

12 Bruce Y. Lee, "Trump States He'll Let RFK Jr. 'Go Wild' on Health, Food and Medicines," *Forbes,* November 2, 2024, updated November 15, 2024,

https://www.forbes.com/sites/brucelee/2024/11/02/trump-states-hell
-let-rfk-jr-go-wild-on-health-food-medicines/.

13 Michael T. Osterholm and Ezekiel J. Emanuel, "How Kennedy Could
    Make It Harder for You and Your Family to Get Vaccinated," *New York
    Times,* November 20, 2024.

INDEX

ABOUT THE AUTHORS

Michael Osterholm is Regents Professor, McKnight Presidential Endowed Chair in Public Health, and the founding director of the Center for Infectious Disease Research and Policy (CIDRAP), all at the University of Minnesota. An internationally renowned epidemiologist and former state epidemiologist of Minnesota, during his fifty years in the field he has led many investigations of outbreaks of international importance.

Dr. Osterholm is the author, with Mark Olshaker, of *Deadliest Enemy: Our War against Killer Germs* and, with John Schwartz, of *Living Terrors: What America Needs to Know to Survive the Coming Bioterrorist Catastrophe*, both of which were *New York Times* bestsellers. The Johns Hopkins Bloomberg School of Public Health named *Deadliest Enemy* the Number One Global Health Book of 2017.

CNN national security analyst and author Peter Bergen wrote, "*Deadliest Enemy* is a lucid and concise account of how the battle against deadly germs is in many ways the most important war of all. *Deadliest Enemy* deftly melds authoritative science with a gripping narrative to tell what is arguably the most important story of our era."

Osterholm appears regularly in the media as a commentator on public health and is the host of the popular *Osterholm Update* podcast, with nearly two hundred episodes. He has authored or co-authored more than 350 papers and abstracts, including twenty-one book chapters. He serves on the editorial boards of six journals and has been an international leader regarding preparedness for an influenza pandemic and has sounded the alarm regarding critical infectious disease threats in *Foreign Affairs,* the *New England Journal of Medicine,* and *Nature,* as well as in the op-ed pages of the *New York Times* and the *Washington Post.* He has been an international leader on the growing concern regarding the use of biological agents as catastrophic weapons targeting civilian populations. In that role, he served as a personal adviser to the late king Hussein of Jordan.

He served as a special adviser to HHS Secretary Tommy Thompson. His successor, Secretary Michael Leavitt, appointed Osterholm to the newly established National Science Advisory Board on Biosecurity, and he was a member of President-Elect Joe Biden's Covid-19 Advisory Board. From 2017 through 2019 he was a science envoy for the US State Department. He serves on the Board of Regents of Luther College and is a member of the National Academy of Medicine, the Council of Foreign Relations, and numerous other professional organizations. He and his partner, Fern Peterson, live in Minneapolis.

Mark Olshaker is an Emmy Award–winning documentary filmmaker and a *New York Times* number one bestselling author of five novels and thirteen books of nonfiction. His books with former FBI special agent and profiling pioneer John Douglas, beginning with *Mindhunter*—the basis for the original Netflix series—and continuing up through *The Killer's Shadow* and *When a Killer Calls,* have sold millions of copies, been translated into many languages, and offer a unique and intriguing perspective into behavioral science and criminal investigative analysis.

His scientific and medical writing is represented by his previous book with Michael Osterholm, *Deadliest Enemy: Our War against Killer Germs,* and *Virus Hunter: Thirty Years of Battling Hot Viruses around the World,* with Dr. C.J. Peters, which the *New York Times* placed on its Noteworthy and Recommended lists for the year and which the *New England Journal of Medicine* compared to Paul de Kruif's celebrated *Microbe Hunters,* declaring it "not merely the exhilarating tale of three decades of scientific research. It is also an outspoken, comprehensive analysis of the political and human issues that frontline scientists fighting outbreaks of hemorrhagic fever deal with daily." Olshaker wrote *The Instant Image: Edwin Land and the Polaroid Experience*, the lead chapter of the medical textbook *Forensic Emergency Medicine*, the IMAX film *Stormchasers*, and the PBS programs *What's Killing the Children?*, *Bioterror: Living with a New Reality*, and *Anatomy of a Pandemic*, among numerous others.

Olshaker's highly praised suspense novels include *Einstein's Brain, Unnatural Causes, Blood Race*, and *The Edge*, which *Publishers Weekly* called "a darkly imagined thriller marked by brisk action and a mind-bending denouement."

His writing has appeared in *Foreign Affairs, Fortune, The Guardian*, the *New York Times, Newsday*, the *St. Louis Post-Dispatch, Time*, the *Wall Street Journal*, the *Washington Post, Washingtonian*, and *USA Today*.

He is the former chairman of the Cosmos Club Foundation and serves on the boards of the Norman Mailer Society and the Rod Serling Memorial Foundation. Olshaker and his wife, Carolyn, an attorney, live in Washington, DC.